Exploring the Ethics of Clinical Xenotransplantation

This book provides a comprehensive examination of the ethics of xenotransplantation.

Due to the global shortage of organs for transplantation, there is increasing interest in cross-species transplantation, or xenotransplantation, among researchers and clinicians. Proponents of xenotransplantation envision a future in which transplantable organs can be safely and efficiently grown in genetically modified pigs. This book provides an overview of the major ethical issues involved in xenotransplantation. Xenotransplantation touches on almost every area of bioethics: animal research, public health, patient selection, allocation, global justice, pediatric research, and informed consent. Each chapter provides a nuanced overview of the positions that have been offered on a specific topic and the authors seek to provide a way forward.

Exploring the Ethics of Clinical Xenotransplantation is an essential resource for researchers, students, and clinicians working in bioethics, medical ethics, animal ethics, and public health policy.

Christopher Bobier, PhD, MA, is Associate Professor of Foundational Sciences at Central Michigan University College of Medicine (Mt. Pleasant, Michigan, USA). In addition to xenotransplantation, his research interests include pediatric ethics, research ethics, philosophy of hope, and animal ethics.

Adam Omelianchuk, PhD, MA, is Assistant Professor at the Center for Medical Ethics and Health Policy at Baylor College of Medicine as well as a practicing clinical ethicist (HEC-C) at Houston Methodist Hospital. His work focuses on ethical issues in death determination and transplant medicine.

Daniel J. Hurst, PhD, ThM, MSc, is Director of Medical Professionalism, Ethics, and Humanities, as well as Associate Professor and Interim Chair of the Department of Medical Education & Scholarship at Rowan-Virtua School of Osteopathic Medicine (Stratford, New Jersey, USA). He is co-editor of *Xenotransplantation: Ethical, Regulatory, and Social Aspects* (2023).

Routledge Annals of Bioethics
Series Editors:
Mark J. Cherry
St. Edward's University, USA
Ana Smith Iltis
Wake Forest University, USA

The Limits of Parental Authority
Childhood Wellbeing as a Social Good
Johan Bester

Bloody Bioethics
Why Prohibiting Plasma Compensation Harms Patients and Wrongs Donors
James Stacey Taylor

Agency, Pregnancy and Persons
Essays in Defense of Human Life
Edited by Nicholas Colgrove, Bruce P. Blackshaw, and Daniel Rodger

Voluntary Consent
Theory and Practice
Maximilian Kiener

An Ethics of Clinical Uncertainty
Lessons from the COVID-19 Pandemic
Mary Ann G. Cutter

Medical Ethics in the Catholic Tradition
Perspectives from Philosophy, Theology, Medicine, and the Law
Margaret M. Hogan, Edward M. Hogan, Annique K. Hogan, and Matthew J.D. Hogan

Exploring the Ethics of Clinical Xenotransplantation
Christopher Bobier, Adam Omelianchuk, and Daniel J. Hurst

For more information about this series, please visit: https://www.routledge.com/Routledge-Annals-of-Bioethics/book-series/RAB

Exploring the Ethics of Clinical Xenotransplantation

Christopher Bobier, Adam Omelianchuk, and Daniel J. Hurst

Routledge
Taylor & Francis Group
NEW YORK AND LONDON

First published 2026
by Routledge
605 Third Avenue, New York, NY 10158

and by Routledge
4 Park Square, Milton Park, Abingdon, Oxon, OX14 4RN

Routledge is an imprint of the Taylor & Francis Group, an informa business

ISBN: 978-1-032-75069-9 (hbk)
ISBN: 978-1-032-75068-2 (pbk)
ISBN: 978-1-003-47222-3 (ebk)

DOI: 10.4324/9781003472223

Typeset in Sabon
by Apex CoVantage, LLC

Contents

	Introduction	1
1	Organ transplantation and alternatives—the state of affairs	10
2	Is xenotransplant animal research ethical?	25
3	What are the public health concerns?	42
4	Who should receive a xenotransplant?	58
5	Is informed consent possible?	75
6	What interests are at stake for children?	91
7	How do we balance innovation and price?	105
8	What are the global and future implications?	124
	Index	*140*

Introduction

This book is meant to provide an overview of the major ethical issues involved in solid organ cross-species transplantation, known as xenotransplantation (so termed because, in Greek, *xenos* means "other" or "foreign"). There is increasing interest in the topic of xenotransplantation, especially among researchers and clinicians, because the major barrier to its use—immunological graft rejection—appears to be crumbling in the face of scientific advances. It is not unusual to see xenotransplantation addressed in the world's leading high-impact medical journals, such as *Nature*, *The Lancet*, and *The New England Journal of Medicine* (Eisenson et al., 2024; Shaver and Bacchetta, 2023; Montgomery et al., 2022). Proponents of xenotransplantation envision a future in which transplantable organs can be safely and efficiently grown in genetically modified pigs—those that have been genetically engineered to reduce immunogenicity and increase physiological compatibility with the human recipient—and there is reason to be optimistic that this future is not far off, as initial clinical trials for kidney xenotransplantation have begun (Pullen, 2024; Rabin, 2025). Xenotransplantation, however, touches on almost every area of bioethics: animal research, public health, patient selection, allocation, global justice, pediatric research, and informed consent, to name a few.

To see why, consider the (imaginary) case of Anna, a 62-year-old mother and grandmother, who suffers from end-stage heart failure and continues to experience shortness of breath, abdominal pain, and fluid retention despite adhering to an optimized medication-based treatment plan. After a recent hospitalization and observing her downward trend over the last two years, a clinician informs her that she has probably less than a year to live unless she receives a heart transplant or a left-ventricular-assist device (LVAD). Because of the experience of friends in her patient support group, Anna is hesitant about living with an LVAD. Although an LVAD would likely be beneficial from a cardiology standpoint, Anna fears that the burdens imposed by the device would further diminish the quality of her life as it did in the lives of some of her friends. Fortunately, her condition satisfies

DOI: 10.4324/9781003472223-1

criteria for transplantation, which means she can be listed to receive a heart. Unfortunately for her, however, there are many people also waiting for a heart and, because hearts are hard to come by, she risks deterioration and being taken off the list. In fact, there is a good chance that she will deteriorate and will be eligible for neither a heart nor an LVAD.

Yet the situation is not limited to just two options, and her clinician makes her aware of another possibility. There is a clinical trial enrolling patients with end-stage heart failure for a transplant of a different kind: researchers can replace Anna's failing heart with a genetically engineered porcine heart. The clinician is willing to refer her to the investigators for an evaluation because she appears to meet the inclusion criteria. How should the patient be approached, by the investigators themselves or along with an "independent" clinician? If they were to enroll her, what would be the conditions for doing so? What should Anna be told about what to expect? What sort of benefits, if any, could she experience? What exactly are the risks? Supposing she is enrolled, should she be taken off the list for a human heart? Only if she actually receives a porcine heart? Or can she still be listed for a human heart even if she receives a porcine heart? Does Anna incur any special obligations if she receives a porcine heart? What about her close contacts? Is Anna, not to mention the transplant team, complicit in the exploitation of animals if she chooses to receive one? Is she, somehow, "less human" or has she in some way compromised her "human dignity" in virtue of receiving a porcine heart? Where does this heart come from exactly? Who raised the "donor" animal, and what are the costs associated with receiving it? It is not hard to imagine Anna experiencing a poor outcome and regretting her decision to refuse an LVAD. Her quality of life could be quite a bit worse if things do not turn out well and it is hard to know if they will. But they might. What should researchers, patients, transplant surgeons, advocacy groups, and the general public think of all this?

Raising these questions is easy but answering them is difficult. Any substantive answer to them involves value-laden assumptions and therefore requires careful ethical analysis. Some of these questions have not been as urgent to answer because there have not been any clinical trials to enroll in, but that is now changing. Historically, xenotransplantation has not offered any clinical benefit to potential research subjects. That, however, appears to be changing as well. Nonetheless, some important questions remain to be settled to make high-quality clinical trials a reality (summarized from Fedson et al., 2024):

- Should the transplantation of porcine kidneys be perfected before the transplantation of porcine hearts and other unpaired organs?
- Should more organ-specific data be gleaned from grafts transplanted into "brain-dead" bodies before trials begin in living patients?

- What trial design, consistent with appropriate regulatory oversight, is optimal for maximizing generalizable knowledge?
- How should conflicts of interest between organ suppliers—those who have a financial interest in the companies that raise genetically modified pigs—and patient selection committees be managed?
- What opportunities are there for pediatric patients or socially disadvantaged populations?
- What is the extent of the researchers' obligations to provide follow-up for infectious disease monitoring?

As of this writing in early 2025, there have been six pig-to-human xenotransplants in the United States, two cardiac and four renal (Rubin 2025), some of which were done on the basis of "emergency use" criteria, meaning the recipients had no other option. The xenotransplantation of a liver and kidney into living patients has also taken place in China (Mallapaty, 2024). Multiple researchers have indicated an interest in initiating clinical trials, and the US Food and Drug Administration has just approved two phaseless kidney xenotransplant clinical trials (Rabin, 2025). Therefore, these questions and the questions raised in Anna's case are becoming more urgent to address as even more clinical trials are more likely to begin.

While there has been a year-by-year increase in the number of publications that discuss the ethics of xenotransplantation in peer-reviewed journals, there is no up-to-date monograph offering a sustained or comprehensive examination of the ethics of xenotransplantation. Our book addresses this lacuna by offering an overview of the topic and the whole range of contemporary ethical issues connected with it. The challenge with writing a book on xenotransplantation is that by the time this book is published, there will be more research published, more patents obtained, and more breakthroughs discovered than when we were writing the chapters. Some of the results will be exciting and could not have been foreseen, hopefully for the better; others will be less surprising but still important. Anyone who has been paying attention to the headlines knows that xenotransplantation is a rapidly advancing field. As discussed here and in Chapter 1, there is reason to be optimistic that soon scientists will be able to grow porcine organs that are suitable for human transplantation. The potential clinical utility of xenotransplantation has never looked so promising.

It is important to note, however, that things have not always been so promising. The first heart transplant, if defined in terms of replacing one weakly pumping heart with another, was performed in 1964 by James Hardy who put the heart of a chimpanzee in a 36-year-old male (Hardy et al., 1964). Unlike his celebrated colleague, Christiaan Barnard, who performed the first human-to-human heart transplant in 1967, Hardy became a pariah in the field and is rarely—and perhaps unjustly—referenced positively in the history of medicine, despite being the first to perform a human-to-human lung transplant. Renal transplants from nonhuman primates were also attempted in the early part of the 20th century with dismal results, though they did improve somewhat and showed more promise in the early 1960s, one of which produced nine months of survival and a return to work for one patient (Reemtsma, 1995). The practice was abandoned when dialysis became more accessible and pathways to retrieving healthy organs from "brain-dead" donors became available.

The primary reasons for considering xenotransplantation in those days involved side-stepping problems generated within a much more restrictive era of law. The now-routine practice of taking an organ from a patient who, after being appropriately extubated and allowed to progress to cardiac arrest, was hard to justify. Once the medical team determined that no human organ was forthcoming and they obtained consent for proceeding with the xeno-experiment, the benefits of coordinating the surgery teams with the animal "donor" and the human recipient were evident to everyone involved. These benefits served to justify the controversial "Baby Fae" experiment of 1984, in which a 12-day-old baby girl received a heart from a baboon (Bailey et al., 1985; see Chapter 6). Sadly, the child died from complications associated with acute graft rejection 20 days later. The ensuing media firestorm and voluminous critical commentary seemed to brand xenotransplantation as a scientific dead end at best, and an inductee to a bioethical "hall of shame" at worst.

Yet researchers never gave up on it because the clinical need of those in organ failure always motivates action. That need is further expanded by advances in ever-improving organ supportive therapies that keep patients alive for ever-increasing amounts of time in both the critical and chronic care settings. Indeed, no transplant "waiting list" or a "demand" for organs is possible without these therapies because no one can be listed or express morally significant demands if one is dead or actively dying from organ failure. Furthermore, technical innovations in immunosuppressive therapies, animal husbandry, and genetic engineering have paved the way for procuring functional grafts from mammals that receive much less sympathy from their human counterparts than nonhuman primates. Unlike with nonhuman primates, the instrumentalization of the pig is widely accepted throughout the world and it is hard to imagine why it would be impermissible to use them as organ sources when they are used for food. Given this *prima facie* sense of permission, the allure of xenotransplantation endures despite facing considerable technical, legal, ethical, and cultural challenges to its implementation.

We recognize that xenotransplantation's promise has always been overhyped, and that the oft-quoted joke that its benefits are "right around the corner and always will be" is grimly apt. Even though the results of the pig-to-human transplants are scientifically exciting, the clinical outcomes are, so far, nothing to write home about. The persistent mysteries of the human immune system and the resilience of infectious disease agents in combination with the complex comorbidities of recipients do not inspire a level of confidence commensurate with the expressions of optimism one finds in research proposals and press releases. Still, it is undeniable that the research is progressing.

In any event, this book is not about scientific innovation nor an introduction to the science of xenotransplantation. This is for the better, as we are not scientists but ethicists who have devoted a considerable amount of time to helping the medical community think carefully about the issues posed by xenotransplantation. Fortunately for us, ethical issues remain relatively constant. As readers already know or will soon discover, the ethics of xenotransplantation covers a wide range of perennial issues in ethics, including questions in animal ethics, business ethics, clinical ethics, public health ethics, and more. Kathleen Fenton illustratively calls xenotransplantation a "microcosm of bioethics" (Fenton, 2022). Some issues are constant staples in bioethics. For instance, how do we allocate scarce medical resources such as solid vital organs? What conditions must be met for animal research to be ethical? Some issues are uniquely salient to xenotransplantation. For instance, should xenotransplant recipients have to waive their right to withdraw from long-term monitoring for infectious disease? Does ethical xenotransplant research require "collective" consent, and if so, what would that require?

There is a clear need for a book that introduces readers to the ethics of this cutting-edge technology. The research is progressing at an incredible speed, and as anyone who recalls watching or reading *Jurassic Park* knows, there is a genuine concern that scientists are more interested in accomplishing this feat than whether they should. Yet we don't think scientists are being negligent of the ethical issues. Quite the contrary—our interactions with xenotransplant researchers are encouraging because they care deeply about ethics. Like the pioneers of allotransplantation in the 1960s, they do not want their field to fall into disrepute. That said, there is an important need for further ethical discussion on the issues since the issues involve interests of concern to everyone. Our goal therefore is to introduce readers to a range of ethical issues posed by xenotransplantation. We have opinions on the topics discussed in this book, but as an introduction to the topic, we attempt to provide a fair perspective on differing positions. Moreover, we limit our discussion to solid organ xenotransplantation and leave aside the broader question of mixing animal cells with human cells, because there is greater public interest in solving the organ shortage through xenotransplantation and much of what applies to solid organ transplantation is applicable to islet cell transplantation.

We have organized this book to follow a logical progression from initiation to implementation. We begin by discussing the need for an increase in organs for transplantation and then examine the ethics of using genetically engineered pigs to address the organ shortage. Since this use of pigs raises public health concerns, we then discuss the public health ethical implications of xenotransplantation. Assuming a favorable verdict on these two matters, we turn to address foundational issues in research and clinical practice: who gets to participate in xenotransplantation or receive a genetically engineered organ, and how can we ensure informed consent, even for potential pediatric recipients? It is important to note that xenotransplant research and (hopeful) clinical implementation are sponsored by biotechnology companies, and so, we examine the business ethics of patenting and selling lifesaving organs. We conclude by discussing global and future implications.

In Chapter 1, we discuss the state of organ transplantation. We survey the current options for supporting patients with organ failure and maximizing the organ supply. The ethical issues in allotransplantation are familiar and enduring and set the stage for xenotransplantation. These issues include how to define death and to ethically procure organs, both of which involve negotiating complicated tensions between the interests of donors and the interests of recipients. Debates about the controversies over death criteria, "opt out" versus "opt in" donation models, and financial incentives for donating will be explored within the framework of how to best maximize the organ supply. The allure of xenotransplantation is that it promises to alleviate the burdens on human donors by shifting them to animal sources while maintaining, if not enhancing, the benefits to recipients. The problems of determining death and abiding by the dead donor rule are simply not relevant in xenotransplantation. The supply of transplantable organs could potentially increase and thereby reduce the morbidity and mortality of those on the transplant wait list; allocation decisions might also be expedited more efficiently and fairly if xenotransplants can serve as either "bridge" (a temporary therapy meant to "bridge" a patient to a more permanent therapy) or "destination" therapies depending on their clinical efficacy. The way forward may be to perfect xenotransplantation as one method among many to meet the demand for organs.

In Chapter 2, we engage with the ethics of using animals for organ procurement. The chapter begins by highlighting the novel use of animals in xenotransplantation compared to other uses of animals in scientific research. We then contrast three general approaches to animal moral status. According to a "strong position" of animal moral status, creating and using genetically modified pigs for organ transplantation is simply wrong. According to a "weak position," there is nothing wrong with doing this. According to a "modest position," creating and using genetically modified pigs for organ transplantation may be ethically permissible if it satisfies stringent necessary conditions. The primary focus of this chapter is assessing the claim that this line of research satisfies the necessary conditions for ethical animal research within a "modest position" framework. We highlight points of contention between defenders of xeno-animal research and critics. We discuss the possibility of disenhancing genetically modified pigs such that they are unable to experience emotional and physical pain, which is a novel and controversial suggestion that has recently been posited within xenotransplantation research.

In Chapter 3, we engage the tension between third-party risk and saving human lives via xenotransplantation. We begin with public health ethics and take for granted that researchers have an ethical obligation to ensure public health and to guard third-party individuals from harm resulting from research. How far is the reach of this obligation, though? In general, research that poses significant risk of an epidemic or pandemic is generally unacceptable. Xenotransplant research, however, poses such a risk via zoonotic infection (animal-to-human infection; also known as "xenozoonotic infection" or "xenozoonosis" when considered in the context of xenotransplantation) from the host source. We evaluate three responses to the concern: one that posits the risk to be so severe as to be impermissible on precautionary grounds; one that posits the necessity of the consent of all who are at risk; and one that posits the necessity of xenograft recipients waiving their right to withdraw from biosurveillance. We discuss the strengths and weaknesses of each response.

In Chapter 4, we discuss the question, "Who should receive a xenotransplant?" Current guidelines and scholarly consensus are that research participation should be restricted to seriously ill patients for whom alternative therapies are unavailable. We explain the rationale for this limitation and discuss two lines of criticism: the risk-benefit analysis is too difficult to agree upon and limiting participants to severely ill patients is unfair. Fair subject selection is examined using principles of generalizability, shared burden, fair opportunity, and third-party risk. Risk-benefit concerns are also discussed in this chapter, and we conclude with a sketch of what the clinical application might look like by integrating both therapy and research together in a way that is consistent with ethical norms and the research history of transplantation in general.

In Chapter 5, we examine the complexities involved with informed consent in xenotransplantation. The risk of xenozoonotic disease, coupled with the current recommendation that only patients not eligible for an allotransplantation should be candidates for xenotransplantation, presents difficulties to the informed consent process. Specifically, it appears that would-be recipients are faced with a forced decision to either accept a xenotransplant or accept death. We examine proposed ways of ameliorating this concern, such as the use of Ulysses contracts, and highlight their strengths and weaknesses. The chapter concludes with concrete recommendations for obtaining informed consent.

In Chapter 6, we address what is at stake in pediatrics. Due to the critical lack of healthy, transplantable organs in the pediatric population, along with the few available therapeutic alternatives to cardiac allotransplantation in this population, children with certain conditions may especially benefit from xenotransplantation. Bridging with a genetically engineered pig heart may be lifesaving for pediatric patients. However, pediatric patients cannot consent either to the xenotransplantation procedure or to the attendant posttransplant monitoring that may accompany the procedure for the remainder of the patient's life. Further, performing xenotransplantation in children prior to gathering data on safety and efficacy in adults is controversial. The chapter concludes with a discussion of how to ensure ethical pediatric xenotransplantation.

In Chapter 7, we explore the intersection of business ethics and biomedical ethics. On the one hand, xenotransplantation may be quite lucrative for biotechnology companies that have significantly invested in funding the research and development. It stands to reason, then, that companies may claim patents on organs and/or on genetically modified pigs.

We discuss the ethics of patenting in xenotransplantation, highlighting the case in its favor as well as criticisms. Along the way, we engage the issue of whether and how the US government should and could intervene in patent protections to force biotechnology companies to lower costs for recipients.

In Chapter 8, we explore issues of justice and fairness and what the implications of this could be in the future. For instance, if xenotransplantation demonstrates clinical efficacy, it is likely to, initially, only be available in high-income countries and questions remain on how it will be used and allocated. How will decisions be made about who receives a xenograft or an allograft? Could we end up in a world where the poor receive xenografts, while the rich and privileged have access to allografts, resulting in worsening health disparities? The answers to many of these questions will be contingent on whether xenotransplantation is scalable, where it is available, whether it will be a bridge or destination therapy, and how its outcomes compare to alternative therapies. The chapter concludes with a discussion of the future work that needs attention in regard to the ethics of xenotransplantation.

Answering the questions surrounding the case of Anna will take time. We hope that readers will benefit from our analysis of these topics so that the field of xenotransplantation can proceed with ethical clarity for all those involved.

References

Bailey, L. L. et al. 1985. Baboon-to-human cardiac xenotransplantation in a neonate. *JAMA*, 254(23), 3321–3329. https://doi.org/10.1001/jama.1985.03360230053022.

Eisenson, D. et al. 2024. Consistent survival in consecutive cases of life-supporting porcine kidney xenotransplantation using 10GE source pigs. *Nature Communications*, 15(1), 3361. https://doi.org/10.1038/s41467-024-47679-6.

Fedson, S. et al. 2024. Ethical considerations in xenotransplantation of thoracic organs—a call for a debate on value based decisions. *The Journal of Heart and Lung Transplantation: The Official Publication of the International Society for Heart Transplantation*, 43(7), 1033–1038. https://doi.org/10.1016/j.healun.2024.03.012.

Fenton, K. N. 2022. Xenotransplantation: A microcosm of bioethics. *Artificial Organs*, 46(6), 995–996. https://doi.org/10.1111/aor.14260.

Hardy, J. D. et al. 1964. Heart transplantation in man: Developmental studies and report of a case. *JAMA*, 188, 1132–1140.

Mallapaty, S. 2024. First pig-to-human liver transplant recipient 'doing very well'. *Nature*, 630(8015), 18. https://www.nature.com/articles/d41586-024-01613-4.

Montgomery, R. A. et al. 2022. Results of two cases of pig-to-human kidney xenotransplantation. *The New England Journal of Medicine*, 386(20), 1889–1898. https://doi.org/10.1056/NEJMoa2120238.

Pullen, L. C. 2024. Xenotransplantation moves toward clinical trials. *American Journal of Transplantation: Official Journal of the American Society of Transplantation and the American Society of Transplant Surgeons*, 24(4), 509–511. https://doi.org/10.1016/j.ajt.2024.03.001.

Rabin, R. C. 2025. For patients needing transplants, hope arrives on Tiny Hooves. *The New York Times*, March 10. https://www.nytimes.com/2025/03/10/health/pig-cloning-transplants-kidneys.html.

Reemtsma, K. 1995. Xenotransplantation: A historical perspective. *ILAR Journal*, 37(1), 9–12. https://doi.org/10.1093/ilar.37.1.9.

Shaver, C. M. and Bacchetta, M. 2023. Immune recognition of porcine organs transplanted into humans. *The Lancet*, 402(10408), 1112–1113. https://doi.org/10.1016/S0140-6736(23)01556-8.

1 Organ transplantation and alternatives—the state of affairs

Combining parts from animals and humans to overcome a difficult problem is not a novel idea. Going back more than 3,000 years, it has been most clearly depicted in the Greek myth of Icarus and Daedalus. Daedalus was famously imprisoned with his son, Icarus, by King Minos of Crete. Daedalus devises a plan to escape by using wax and feathers to fashion "bird" wings for him and his son to escape, which, as we know, ended in tragedy. While attempts at xenotransplantation in more recent history have yet to yield the results researchers have hoped for, the future is beginning to look more promising. However, before moving on to focus specifically on xenotransplantation, it is important to understand the contexts and problems that have driven scientists and clinicians toward consideration of cross-species transplantation. In this chapter, we outline the nature of the organ shortage, the ethical and psychological constraints involved with organ donation in general, certain policy proposals for addressing the shortage, future technology solutions, and the role we think xenotransplantation will play in addressing morbidity and mortality associated with organ failure.

The chronic need for organs

Discussions of organ transplantation tend to start by noting two enduring facts: organ transplant is the "gold standard" for treating patients with organ failure, and the barrier to receiving an organ for transplant is its ever-increasing scarcity. For example, in the United States, over 100,000 people are officially listed as waiting for an organ, yet there are only (on average) 40,000 transplants per year (between 2018 and 2022); approximately, 17 people die each day waiting for an organ transplant. Discussions then move from these grim statistics to various proposals for closing the gap between the "supply" and "demand" for organs. Our discussion will largely follow this script, surveying the various proposals for maximizing the number of allografts to set the stage for xenotransplantation's potential

DOI: 10.4324/9781003472223-2

role in this macabre economy, but we depart from it in one crucial respect: we do not assume that the ultimate goal is to "close the gap" between demand and supply and reduce the organ shortage, but rather to reduce the morbidity and mortality associated with waiting for vital organs. As we suggest, there is reason to believe the "gap" will never be closed.

Although the number of transplants done per year in the United States has steadily increased in the last 30 years, the number of waitlist additions per year continually outpaces this growth (see Figure 1.1).

Waitlist additions are supply-sensitive in the sense that more patients are added to the waitlist as the capacity for transplant increases. To be sure, the disease burden of chronic conditions is also increasing (Koye et al., 2018). While one might reasonably think the demand for organs has gone up because kidney health has gone down, the reason kidney health has gone down is not necessarily that we were healthier in the past. Diagnostic methods have improved and the total patient population continues to increase with people living longer and therefore the incidence and prevalence of chronic kidney disease have increased (Chesnaye et al., 2024). Public health initiatives and advances in communicable disease prevention have helped increase the average lifespan throughout the world, which means the incidence and prevalence of the ailments that affect human beings later in life have also grown over time. Waitlist additions will likely continue to grow insofar as our organ supply will grow for the foreseeable future.

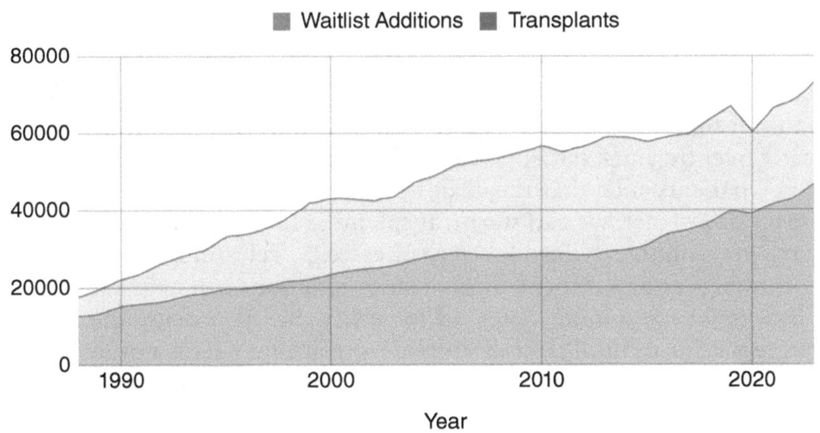

Figure 1.1 Waitlist Additions and Transplants.

Source: Data was collected from OPTN ranging from 1988 to 2023 (https://insights.unos.org/OPTN-metrics/)

Therefore, proposals framed as "potential solutions to the organ shortage" risk producing the opposite effect, creating more demand—a kind of "if you build it, they will come" effect that has been observed in the construction of hospitals with greater bed capacity (Wennberg, 2007). This is because the idea of "need" in organ transplant is constructed from criteria that measure the level of illness one has from a predefined concept of "organ failure" that warrants listing but is not so high that one will not meaningfully benefit from receiving one—a kind of "Goldilocks" zone of illness that "indicates" transplant. What defines "meaningful benefit" depends on the goals of care, which can be measured in terms of discharge from the hospital, release from medical technologies, years-long survivability, return to work or one's daily activities for some meaningful length of time, or prevention of expected morbidity. The more organs there are, the more candidates for transplantation there can be.

Moreover, the so-called "borderline" patients—those who come close, but who are nonetheless excluded from listing—are not necessarily excluded because they do not pass some fixed "bright line" of medical eligibility. Such patients who have serious though manageable secondary medical problems (e.g., diabetes and vascular disease) or need to be re-transplanted because of chronic rejection or have a marginal social support network could benefit from an organ if there are enough to go around. Transplant teams keep such patients off the list in order to maximize the benefits of the few organs they have. Yet as David Cooper and Robert Lanza explain, "If the number of organs becoming available annually were unlimited, *all* of these borderline patients could be given the opportunity of a transplant despite the risk that they may not do as well as others" (Cooper and Lanza, 2000, p. 19). Of course, there will always be a limit to the number of organs available, but it stands to reason that the more that become available, the more the eligibility criteria for transplant expand. The logic goes like this: if the supply of organs increases such that we are able to meet our "clearly over the line" patients, we can start meeting the needs of our "borderline" patients such that they can be designated as "over the line"; once those needs are met, we can start transplanting to meet preventative needs. Today, preventative "lifestyle" measures (e.g., refraining from smoking and drinking) and treatments for chronic conditions can help keep one off the list, but the resultant increased longevity may just delay the need for organ replacement until later in life. Innovations may screen organ-specific cancers before they manifest, indicating the need for a transplant before cancer-related organ failure occurs—assuming there is a large supply of organs available for transplant, of course (Cascalho and Platt, 2005). The point is that "need" is defined by the benefits sought under a given set of circumstances. So, as long as physicians and patients seek benefit through organ replacement, there will always be a "need" or a "demand" that will put pressure on our stock of supply even if it becomes more plentiful.

Placement on the waitlist is also a function of improved therapies that "bridge" a patient to the "destination" of transplant. We know that the very existence of the list and the phenomenon of "demand" depend on the availability of therapies that sub-optimally replace or supplement organ function, or minimize the dysfunction of a failing organ (Beard et al., 2013). Ironically, the relative effectiveness of these therapies helps create demand. The reason there is such a large waiting list for kidneys is that dialysis is both widely available and fairly effective in replacing kidney function that keeps the patient alive in the chronic care setting. The ever-improving efficacy of temporary cardiology interventions such as extracorporeal membrane oxygenation (ECMO), Impellas, left ventricular assist device (LVAD), and total artificial hearts will have a similar effect, at least in the critical care setting as waiting times on the lists for heart transplants can be extended. Even if these therapies, like LVADs, become so effective that they can be "destinations" for some patients, organ transplant will likely still be considered the "gold standard" for maximizing medical and quality-of-life benefits. Hence, it is unclear if there will ever be a substantive decrease in the gap between "supply" and "demand" for organs.

In short, then, the goal of solving the "organ shortage" is misleading. As much as that matters as a *means* to an end, the end goal should be to *reduce the morbidity and mortality* of those on the waitlist. It is those people who would *benefit most* from receiving an organ. In the United States, the number of people who have died on the waitlist has hovered between 7,000 and 8,000 per year for most of the past 25 years, though the numbers were trending downward before the COVID-19 pandemic; 2023 is the first year to reach levels measured in 1999 (see Figure 1.2). What accounts for this is not entirely clear, but of note is the growing number of organs procured from deceased donors (Figure 1.3). In any event, proposals aimed at reducing waitlist mortality are needed.

Proposals for closing the gap

Various proposals for "closing the gap" include revising consent arrangements, implementing market forces, or rethinking the ethics surrounding death, dying, and the dead donor rule.

Allotransplantation, whether from living or deceased donors, already involves several ethical issues, some of which are difficult to manage even if there are compelling approaches for settling them. For example, living donation involves an invasive workup and surgery that provides no benefit and risks harm to the donor's body, which appears to conflict with

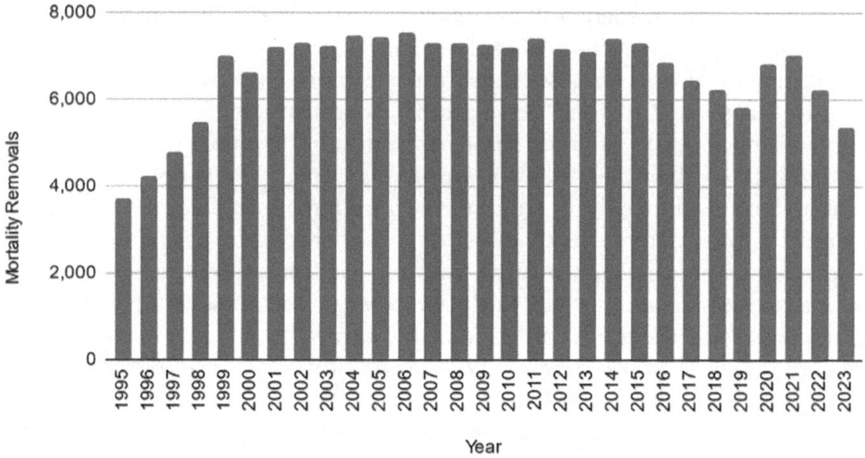

Figure 1.2 Removals From the Waitlist Due to Mortality.

Source: Data was collected from OPTN 1995 to 2023. Search by visiting https://optn.transplant.hrsa.gov/data/view-data-reports/national-data/# and selecting "Death Removals by Region by Year"

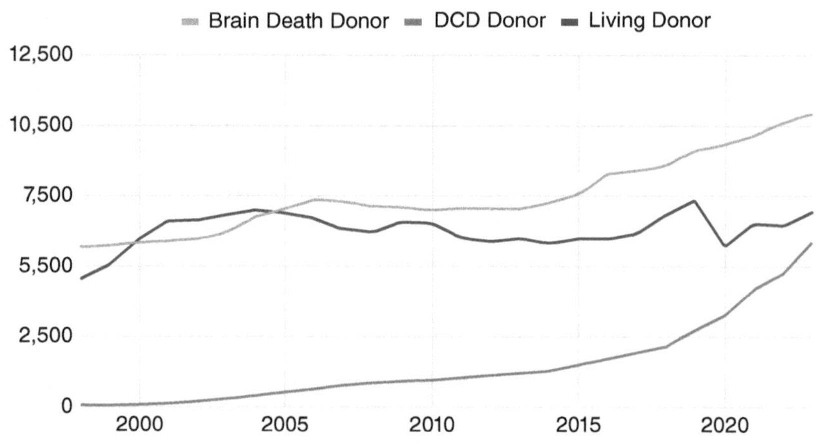

Figure 1.3 Brain Death Donor, DCD Donor, and Living Donor.

Source: Data was collected from OPTN 1998 to 2023. Search by visiting https://optn.transplant.hrsa.gov/data/view-data-reports/national-data/#. Choose the category "Donor" and choose "All Organs." Then, choose "All Donors by Donor Type." After the screen changes, select "Add field to report: Donation after Circulatory Death (DCD)."

the time-honored rule against introducing medically caused harms into the body (Murray, 1964). Resolving this conflict through a broader concept of "benefit" balanced with a complicated risk assessment, controlled through the mechanism of informed consent involves a process (not a "signature" event) that must involve safeguards against deception, coercion, and undue influence that provides ample opportunities for the donor to withdraw from the process entirely.

Insofar as deceased donation goes, it depends on there being an identifiable condition of the body that is in bad enough shape to declare it dead but in good enough shape for it to supply functional vital organs that would be effective for transplant. While the conditions identified in terms of "brain death" or the "permanent" (but not technically irreversible) cessation of circulatory and respiratory functions are widely accepted by medical and legal establishments around the world, they do not comport with a "commonsense" or "folk" idea of death, which is popularly associated with coldness, paleness, and stiffness. The general lack of awareness that potential donors have of these facts creates problems for the authorization process, since the information that would matter to the decision to donate is neither well-disclosed nor well-understood. These problems, which are already hard and contentious, are only exacerbated when the gap between supply and demand for organs increases, and researchers look for novel ways to close it.

Proposals to expand living donation include more or less extreme measures. Some have argued that we should allow organs to be bought and sold, permitting market forces to incentivize increasing the supply (Hippen, 2006; Beard et al., 2013). Others, more modestly, propose financially supporting donors for their time, lost wages, and medical expenses through some specialized insurance-style provider or government-funded health authority (Gershun and Lantos, 2021; Rodger and Venter, 2023). Examples of the latter already exist in the United Kingdom, where living kidney donors can apply for financial remuneration from the National Health Service, though the financial impact is still considered problematic by prospective donors.

Some think the class of patients who are "imminently dying" should be permitted to donate their kidneys or liver, then be terminally extubated and given comfort care after their donation (Morrissey, 2016; Ackah et al., 2017). The most extreme proposals (entertained only by philosophers) involve "lotteries" in which organs are conscripted from healthy members of the general public regardless of the wishes of those who are made to supply them (Harris, 1975).

Proposals to expand deceased donation also involve revising the definition of death. Historically, transplantation interests motivated the construction of "brain death" criteria, which, though widely accepted around the world, still cause unease and dissent within the clinical encounter, and critical scrutiny from medically trained critics (Ad Hoc Committee of the Harvard Medical School, 1968; Shewmon, 2021; Eble et al., 2024). Nonetheless, this "neurocentric" view of death combined with organ donation interests has motivated the construction of so-called "unified brain theories" of death, which define death in terms of the loss of meaningful brain circulation, and then stipulate two methods for determining that loss, either by the standard neurologic examination used in "brain death" determinations or by the traditional circulatory-respiratory criteria (Bernat et al., 2023; Bernat, 2024).

Those who hesitate to further revise the definition of death sometimes reject the rule that transplant teams should not be involved in the death of the donor and that something like "organ donation euthanasia" or "double effect donation" should be permitted (Wilkinson and Savulescu, 2012; Camosy and Vukov, 2021). In these views, what gives life its value is separable from our death concepts, making death irrelevant to the criteria for unpaired organ donation (e.g., liver, heart, or pancreas). To be a donor, one must fall below a certain quality-of-life threshold that makes lethal surgery permissible to choose. Thus, the substitution of the "dead donor rule" for an "as good as dead donor rule" is thought to provide a path forward to expand the donor pool by expanding the number of bodies that meet the relevant eligibility criteria (Omelianchuk, 2022). Yet proponents of such a view admit that such a change to transplant practice risks creating public distrust and a net loss of donors willing to donate (Miller and Truog, 2008).

Donation after a circulatory determination of death (DCDD) has produced a substantial increase of organs for transplant in recent years and policymakers suggest expanding the conditions under which it can be done (Robert, 2007). The majority of DCDD procurements throughout the world, and exclusively in the United States, occur in the "controlled" setting where the patient is intubated in the ICU with a devastating brain injury (but isn't "brain dead") and a decision has been made to extubate and allow natural death. Once that decision is made, the question of organ donation is settled. If it is settled in the favor of donation, the patient is extubated in the operating room and must progress to mechanical asystole in under 90 minutes (at most). Once that occurs, a five (though sometimes two) minute "no touch" period is observed to rule out autoresuscitation, and no other contraindications of death are observed. When that time has passed, the donor is declared dead by a member of the treating team. Only then can the transplant team start procurement. By contrast, the "uncontrolled" variety retrieves organs from those whose deaths are unexpected, regardless of whether they are in the hospital, and organ preservation measures begin only after all life-sustaining efforts have failed. While some think that implementing "uncontrolled" DCDD would lead to an increase in donation opportunities (as many as 22,000 per year in the United States), the chief problem it faces is determining consent to authorize donation under such variable circumstances (Wall et al., 2015).

Others suggest that consent to donate should be "presumed" unless one has previously "opted out" (Kennedy et al., 1998). The evidence supporting this proposal is mixed (Ahmad et al., 2019; Arshad et al., 2019). Spain has shown an increase in organs for transplant, while the UK, probably because of COVID-19, has not (Streit et al., 2023; Ugur and Molina Pérez, 2023); donation rates actually decreased in Chile (Domínguez and Rojas, 2013). The policy falls out of the "choice architecture" literature that assumes that it is permissible to create policies that take advantage of the heuristics and biases inherent to human cognition in order to "nudge" people to choose a pro-social option in a way that is consistent with their autonomy (Thaler and Sunstein, 2009). While provocative, and perhaps promising, evidence of the "opt out" policy's utility is mixed at best as it depends on how sensitive procurement agents are to "family veto"—the phenomenon of procurement being refused on the word of the family rather than the patient.

Others, seeking to remain within currently accepted frameworks for DCDD protocols, believe that reperfusing the donor's organs in situ with the aid of bypass technologies and restarting the donor's heart to reverse warm ischemic damage is consistent with a circulatory determination of death so long as the surgeons take measures to occlude blood flow to the brain (Wall et al., 2024). While the technique shows very promising clinical utility, it is far from clear that it complies with the dead donor rule or legal requirements of an "irreversible" or "permanent" cessation of circulation to determine death (Omelianchuk et al., 2024). Nor is it clear that there is 100% certainty that occluding the vessels blocks sufficient flow to the brain to ensure no circulation to the brain, especially from collateral vessels, continues (Manara et al., 2020; Couzin-Frankel, 2023). Extending the "no touch" time from 5 minutes to 20 might alleviate worries about the dying process being interfered with (Gerosa et al., 2024), but doing so would likely not contribute much in the way of maximizing the organ yield.

One promising technology that does not disrupt existing frameworks is the use of ex vivo machine perfusion methods that can "pump" organs after explant and before transplant (López-Martínez et al., 2024). The technology is expensive and requires highly trained staff to use effectively, but the results are very promising. The techniques have allowed for "expanded criteria" for donation—taking a marginal graft from a donor that would otherwise risk being discarded and reconditioning for better results (Malkawi et al., 2023). Extending time on the pump allows for more time to make wise allocation decisions and reduce waste (Groen et al., 2023). Comparable outcomes with brain-dead donors or donors who are subject to in situ normothermic regional perfusion have been reported (Mohkam et al., 2022; Schroder et al., 2023).

While some of these proposals are more reasonable than others, most of them would require a (radical) shift in our understanding of transplant ethics to fully accommodate them. In any event, we are left with the impression that almost anything is better than the status quo. So great is the imperative to maximize the organ supply that no time-honored belief in medical ethics, whether it be concern for bodily integrity, informed consent, economic fairness, the wrongness of killing, or even the nature of life and death, is safe from critical scrutiny. Hence, xenotransplantation's appeal is that it would (1) allow us to maintain current allotransplantation practices, which are widely seen as beneficial and worthy of safeguarding and commending to the general public, and (2) substantially lessen the gap between supply and demand, something allotransplantation alone does not seem capable of doing even if many of the proposed changes suggested earlier are implemented.

The ethical promise and limits of xenotransplantation

Shifting some of the burden of the organ demand to nonhuman animals could help avoid some of the controversies bound up in allotransplant proposals designed to maximize the organ supply. That is to say, xenotransplantation could help remedy the organ shortfall while keeping allotransplantation consistent with the following long-standing (if "conservative") values that govern allotransplantation in the United States (gathered from Crowe and Cohen, 2006):

- The promotion of the goods of solidarity, friendship, and health that saves lives facilitated by a generosity-based public policy, and not the corrupting forces of markets.
- Safeguarding the human activity of donation as a personal decision requiring voluntary informed consent, something that cannot simply be "presumed" or "implied."
- Ensuring respect for the loved ones of deceased donors, especially during the phase beginning with the decision to withdraw life-sustaining treatment and ending with the early stages of mourning.
- Upholding the rule that in the case of deceased donors, transplant surgery ought to only begin once the donor is soundly declared dead.
- The fair application of principles that distribute organs equitably and on the basis of clinical utility.

The point is not to suggest that allotransplantation must stay consistent with these values or that other proposals involving "presumed consent" are automatically out of bounds, but that the proposed changes may not be necessary and that the status quo policies for allotransplantation can stay in place. The more organs there are, the less aggressive allotransplant needs to be in finding them.

On the other hand, xenotransplantation will likely be disruptive to the status quo and met with hesitancy during the time it takes to become perfected and beyond. This hesitancy may stem from animal welfare concerns, a standing preference to receive only an allograft, a distrust of genetic engineering, or a distrust of the businesses and institutions that furnish the xenografts (or some combination). Thus, the proposals for changing the allotransplant system may still be warranted to some extent despite xenotransplantation's contribution to the organ supply. Whatever is decided, however, the goal of shifting the burden from humans to animals should be to reduce morbidity and mortality of those on the waiting list.

Shifting the burden depends, of course, on xenotransplantation becoming a safe, effective, and affordable option, which requires prioritizing its research and development. The goal is to create a source of organs with comparable outcomes to the "gold standard" of allotransplantation. The ethics of how to do this will be explored more in each chapter, but it will be instructive to reflect on the limits of xenotransplantation even if it becomes capable of reaching "gold standard" status.

Imagining total replacement

The question to be answered is this: What place, if any, would xenotransplantation have in meeting the needs of patients in need of an organ transplant? The question is hard to answer and to illustrate why, we can start by imagining xenotransplantation in a perfected state—one that gives it a dominant and exclusive place in meeting transplant needs—to highlight the relevant complexities. In the perfected state, xenotransplantation is fully capable of not only replacing allotransplantation but also going beyond it, meeting every demand for organ transplant. Yet this scenario of "total replacement" depends on (1) the ethical acceptability of raising and culling animals exclusively for their organs (the animal-use condition); (2) the unacceptability of being listed only for human-to-human transplants (the "no same-species preference" condition); and (3) there being an efficient and sustainable system of distribution capable of delivering grafts to candidate recipients (the easy distribution condition). Each of these conditions, the themes of which we discuss in other chapters, faces enduring objections from animal welfare, patient autonomy, and economic feasibility arguments that, when taken together, are potentially devastating to the proposal of total replacement.

Even if xenotransplantation were to attain the capabilities for the perfected state, there would still be a large segment of the patient population who would either object to the animal-use condition or retain the same-species preference in their medical decision-making, and yet still be eligible for allotransplants that come from living or deceased donors (as mentioned earlier). Moreover, there would likely still be a preference to donate one's organs that is hard to reject as morally objectionable. It is hard to imagine a world in which a transplant of a single kidney between identical twins would be deemed unethical, or if not unethical, then gratuitous—a funny case of extravagance that would invoke nostalgia and perhaps pity for the sort of people Joseph Murray aided in 1954 when he performed the first successful kidney transplant between the Harrick brothers (Murray et al., 1955). Indeed, such an action is *laudable* because it promotes the common and individual goods of solidarity, friendship, and health through the virtues of benevolence, generosity, and the acknowledgment that we are ultimately dependent on one another to flourish as individuals. With respect to animals, xenotransplantation already treads upon their welfare for the sake of promoting and enhancing human welfare. Yet the perfected state would only expand their use and entrench the attitude that human needs are to be met, whenever possible, at the expense of animals, always and everywhere—an attitude that can hardly be justified if only because it recognizes nothing in animals that deserves respect. So, the perfected state's proposal of total replacement crowds out laudable actions and makes room for dubious, if not vicious, attitudes.

To be sure, the perfected state of total replacement may not be realistic. It seems plausible that xenotransplantation, at best, will only play a role in supplementing allotransplantation in closing the gap between "demand" and "supply" of organs for transplant. Varying attitudes about respect for animals and the preference of receiving only a human organ will likely endure, as will the spirit of generosity that donors already exhibit in their efforts to undergo living or deceased donation. Yet there is enough acceptance of the animal-use condition and probably enough indifference to the values underlying the same-species preference (though this needs more research) to make xenotransplantation an attractive option and to research ways that would ultimately make it a safe, effective, and affordable therapy. Nonetheless, questions remain to be settled about justice in the distribution of organs. It is not wholly unrealistic to imagine a scenario where low-income individuals are more likely to become recipients of xenografts, leaving higher income individuals more likely to receive a human organ. Where the outcomes of an allotransplant remain superior, we risk compounding existing health inequalities.

Beyond xenotransplantation?

Ideally, we would be able to obviate all the ethical problems, and perhaps the medical problem of immunosuppression, found in both allotransplantation and xenotransplantation if we were able to produce synthetic organs sourced from the recipient's genetic constitution. To make that dream a reality, funding the research of synthetic organs—mechanical or organically bioprinted—should be a higher priority than it is, especially if avoiding the ethical and medical problems associated with using organs from human and animal bodies is most consistent with our values. This is the most interesting objection to xenotransplantation research, because it sets a high ethical standard that is even hard for allotransplantation to meet. Since allotransplantation is more justifiable than xenotransplantation, the priority of funding the research and development of xenotransplantation is further diminished, leaving some to question whether xenotransplantation should even be considered given its unique and perhaps even existential risks, specifically a pandemic spread of zoonotic infectious disease (see Chapter 3).

The production of synthetic organs can boast of some exciting progress, but the day when they could be considered a safe and effective alternative to allotransplants is still a long way off. It must be admitted that predictions about the potential use of the technologies involved are not easy to make, but it seems plausible that the most likely way they will help reduce the need for transplants is by repairing defects in the patient's failing organs and not by totally replacing them. Whether or not such regenerative medicine techniques will make things easier for people suffering from organ failure will depend on how accessible and affordable that sort of technology is.

In any event, the most popular idea in the scientific imagination these days is using the power of 3D bioprinting to produce organs for transplant (Vijayavenkataraman et al., 2018). 3D bioprinting depends upon scanability (inputting the organ-to-be replicated), translatability (modeling the input information into an adequate representation of the desired organ), and printability (outputting that representation into a serviceable organ). Each part of the process is complex and difficult to do even with simple tissues. While advances have been made in bioprinting cartilage (which has been shaped into the outer part of the ear), much more work needs to be done to efficiently manipulate stem cells into differentiating the needed tissues for producing the architectural and vascular complexity of human organs. This needs to happen before any measure of function or durability could be tested and retested to the point that they could be a viable, if expensive, alternative to allotransplants (Datta et al., 2023). The costs of such a research program are not only financial, they also include the ongoing morbidity and mortality of those waiting for a transplant while this technology develops. If we make no room for xenotransplantation to meet our transplant needs, these costs have to be accepted while synthetic solutions are developed. Yet it is far from obvious that we should accept these costs.

By contrast, xenotransplantation's promise of providing safe and effective xenografts for transplant is not as far off, or at least does not seem to be, given the current state of the preclinical science. This can be seen by the several xenotransplants permitted via the expanded access pathway in the United States, as well as those performed in China. The data gathered from these studies has helped to inform and support the start of first-in-human kidney xenotransplantation clinical trials in the United States in 2025. However, its success will depend on whether the immune rejection problem can be overcome, the availability of genetically modified pigs, and whether the risks of cross-species infectious disease can be sufficiently minimized. However, those issues are solvable through improved medicine and methods. One issue that cannot be solved by those means is whether to "shift the burden" to animals in the first place. What reasons, if any, could justify the animal-use condition, especially in research that may not develop into a validated therapy? What are the arguments against using animals in such research? Beyond research, why think a reduction of mortality and morbidity for those on the waitlist is worth the cost to the animals involved? These questions and others like them will be considered in the next chapter.

References

Ackah, R. L., Rana, A. and Goss, J. A. 2017. Imminent death liver donation for children: Potential to meet an unmet need. *JAMA Surgery* [Preprint]. https://doi.org/10.1001/jamasurg.2017.3793.

Ad Hoc Committee of the Harvard Medical School. 1968. A definition of irreversible coma: Report of the ad hoc committee of the Harvard Medical School to examine the definition of brain death. *JAMA*, 205(6), 337–340. https://doi.org/10.1001/jama.1968.03140320031009.

Ahmad, M. U. et al. 2019. A systematic review of opt-out versus opt-in consent on deceased organ donation and transplantation (2006–2016). *World Journal of Surgery*, 43(12), 3161–3171. https://doi.org/10.1007/s00268-019-05118-4.

Arshad, A., Anderson, B. and Sharif, A. 2019. Comparison of organ donation and transplantation rates between opt-out and opt-in systems. *Kidney International*, 95(6), 1453–1460. https://doi.org/10.1016/j.kint.2019.01.036.

Beard, T. R., Kaserman, D. L. and Osterkamp, R. 2013. *The Global Organ Shortage: Economic Causes, Human Consequences, Policy Responses*. Stanford University Press. https://doi.org/10.11126/stanford/9780804784092.001.0001.

Bernat, J. L. 2024. The unified brain-based determination of death conceptually justifies death determination in DCDD and NRP protocols. *The American Journal of Bioethics*, 24(6), 4–15. https://doi.org/10.1080/15265161.2024.2337392.

Bernat, J. L. et al. 2023. Understanding the brain-based determination of death when organ recovery is performed with DCDD in situ normothermic regional

perfusion. *Transplantation*, 107(8), 1650–1654. https://doi.org/10.1097/TP.0000000000004642.
Camosy, C. C. and Vukov, J. 2021. Double effect donation. *The Linacre Quarterly*, 88(2), 149–162. https://doi.org/10.1177/0024363921989477.
Cascalho, M. and Platt, J. L. 2005. New technologies for organ replacement and augmentation. *Mayo Clinic Proceedings*, 80(3), 370–378. https://doi.org/10.4065/80.3.370.
Chesnaye, N. C., Ortiz, A., Zoccali, C., Stel, V. S. and Jager, K. J. 2024. The impact of population ageing on the burden of chronic kidney disease. *Nature Reviews Nephrology*, 1–17.
Cooper, D. K. C. and Lanza, R. P. 2000. *Xeno: The Promise of Transplanting Animal Organs into Humans*. Oxford and New York: Oxford University Press. http://site.ebrary.com/id/10278247 (Accessed: 7 June 2024).
Couzin-Frankel, J. 2023. Giving heart. *Science*, 380(6645), 572–576.
Crowe, S. and Cohen, E. 2006. *The President's Council on Bioethics: Organ Transplantation: Potential Policy Recommendations*. https://bioethicsarchive.georgetown.edu/pcbe/background/organ_recs.html (Accessed: 8 June 2024).
Datta, P., Cabrera, L. Y. and Ozbolat, I. T. 2023. Ethical challenges with 3D bioprinted tissues and organs. *Trends in Biotechnology*, 41(1), 6–9. https://doi.org/10.1016/j.tibtech.2022.08.012.
Domínguez, J. and Rojas, J. L. 2013. Presumed consent legislation failed to improve organ donation in Chile. *Transplantation Proceedings*, 45(4), 1316–1317. https://doi.org/10.1016/j.transproceed.2013.01.008.
Eble, J. M., Di Camillo, J. A. and Colosi, P. J. 2024. Catholics united on brain death and organ donation: A call to action. *Ethics & Medics*, 49(3), 1–4. https://doi.org/10.5840/em20244935.
Gerosa, G. et al. 2024. Twenty-minute no-touch period before controlled DCD heart retrieval: Is there still a chance for successful recovery? The Italian experience. *The Journal of Heart and Lung Transplantation*, 43(4), S257. https://doi.org/10.1016/j.healun.2024.02.1165.
Gershun, M. and Lantos, J. D. 2021. *Kidney to Share*. Ithaca and New York: ILR Press.
Groen, P. C., de Jonge, J. and Porte, R. J. 2023. Prolonged normothermic machine perfusion: Buying more time for liver graft assessment and repair. *Transplantation*, 107(6), 1221–1222. https://doi.org/10.1097/TP.0000000000004553.
Harris, J. 1975. The survival lottery. *Philosophy*, 50(191), 81–87.
Hippen, B. 2006. The case for kidney markets. *The New Atlantis*, 14, 47–61.
Kennedy, I. et al. 1998. The case for "presumed consent" in organ donation. *The Lancet*, 351(9116), 1650–1652. https://doi.org/10.1016/S0140-6736(97)08212-3.
Koye, D. N., Magliano, D. J., Nelson, R. G. and Pavkov, M. E. 2018. The global epidemiology of diabetes and kidney disease. *Advances in Chronic Kidney Disease*, 25(2), 121–132.
López-Martínez, S., Simón, C. and Santamaria, X. 2024. Normothermic machine perfusion systems: Where do we go from here? *Transplantation*, 108(1), 22–44. https://doi.org/10.1097/TP.0000000000004573.
Malkawi, D. et al. 2023. The role of normothermic machine perfusion in extended criteria donor grafts: A new direction in liver graft assessment and preservation. *Livers*, 3(4), 709–726. https://doi.org/10.3390/livers3040046.
Manara, A. et al. 2020. Maintaining the permanence principle for death during in situ normothermic regional perfusion for donation after circulatory death organ recovery: A United Kingdom and Canadian proposal. *American Journal of Transplantation*, 20(8), 2017–2025. https://doi.org/10.1111/ajt.15775.

Miller, F. G. and Truog, R. D. 2008. Rethinking the ethics of vital organ donations. *Hastings Center Report*, 38(6), 38–46.

Mohkam, K. et al. 2022. In situ normothermic regional perfusion versus ex situ normothermic machine perfusion in liver transplantation from donation after circulatory death. *Liver Transplantation*, 28(11), 1716–1725. https://doi.org/10.1002/lt.26522.

Morrissey, P. 2016. Kidney donation before imminent circulatory death. *American Journal of Kidney Diseases*, 68(4), 515–517. https://doi.org/10.1053/j.ajkd.2016.04.013.

Murray, J. E. 1964. Moral and ethical reflection on human organ transplantation. *Linacre Quarterly*, 31, 54–56.

Murray, J. E., Harrison, J. H. and Merrill, J. P. 1955. Renal homotransplantation in identical twins. *Surgical Forum*, 6, 432–436.

Omelianchuk, A. 2022. The inviolateness of life and equal protection: A defense of the dead-donor rule. *Theoretical Medicine and Bioethics*, 43(1), 1–27. https://doi.org/10.1007/s11017-022-09557-4.

Omelianchuk, A. et al. 2024. Neither ethical nor prudent: Why not to choose normothermic regional perfusion. *Hastings Center Report* [Preprint]. https://doi.org/10.1002/hast.1584.

Robert, S. 2007. Organ donation after cardiac death. *New England Journal of Medicine*, 357(3), 209–213. https://doi.org/10.1056/NEJMp078066.

Rodger, D. and Venter, B. 2023. A fair exchange: Why living kidney donors in England should be financially compensated. *Medicine, Health Care, and Philosophy*, 26(4), 625–634. https://doi.org/10.1007/s11019-023-10171-x.

Schroder, J. N. et al. 2023. Transplantation outcomes with donor hearts after circulatory death. *New England Journal of Medicine*, 388(23), 2121–2131. https://doi.org/10.1056/NEJMoa2212438.

Shewmon, D. A. 2021. Statement in support of revising the uniform determination of death act and in opposition to a proposed revision. *The Journal of Medicine and Philosophy* [Preprint]. https://doi.org/10.1093/jmp/jhab014.

Streit, S. et al. 2023. Ten lessons from the Spanish model of organ donation and transplantation. *Transplant International*, 36, 11009. https://doi.org/10.3389/ti.2023.11009.

Thaler, R. H. and Sunstein, C. R. 2009. *Nudge: Improving Decisions About Health, Wealth, and Happiness*. Revised and Expanded edition. New York: Penguin Books.

Ugur, Z. B. and Molina Pérez, A. 2023. The toll of COVID-19 on organ donation and kidney transplantation in Europe: Do legislative defaults matter? *Health Policy*, 136, 104890. https://doi.org/10.1016/j.healthpol.2023.104890.

Vijayavenkataraman, S. et al. 2018. 3D bioprinting of tissues and organs for regenerative medicine. *Advanced Drug Delivery Reviews*, 132, 296–332. https://doi.org/10.1016/j.addr.2018.07.004.

Wall, A. E. et al. 2024. The American society of transplant surgeons consensus statement on normothermic regional perfusion. *Transplantation*, 108(2), 312–318. https://doi.org/10.1097/TP.0000000000004894.

Wall, S. P., Plunkett, C. and Caplan, A. 2015. A potential solution to the shortage of solid organs for transplantation. *JAMA*, 313(23), 2321–2322. https://doi.org/10.1001/jama.2015.5328.

Wennberg, J. E. 2007. *Supply-Sensitive Care: A Dartmouth Atlas Project Topic Brief*. The Dartmouth Institute for Health Policy and Clinical Practice. http://www.ncbi.nlm.nih.gov/books/NBK586632/ (Accessed: 31 May 2024).

Wilkinson, D. and Savulescu, J. 2012. Should we allow organ donation euthanasia? Alternatives for maximizing the number and quality of organs for transplantation. *Bioethics*, 26(1), 32–48. https://doi.org/10.1111/j.1467-8519.2010.01811.x.

2 Is xenotransplant animal research ethical?

In 2009, Jeff Blumenfeld, a loving father, grandfather, and husband from Colorado, was diagnosed with kidney cancer, which resulted in one of his kidneys being removed. Then, beginning in 2022, his second kidney started to fail. He was immediately placed on the kidney transplant list as well as on dialysis to remove extra fluid and waste products from his blood (Allen, 2023). Dialysis is not curative, however, and it can be costly, painful, and time-consuming. The best treatment Jeff could hope for was a kidney transplant. Unfortunately, for those in need of a kidney transplant, the wait is long and uncertain: 36 people died in Colorado in 2022 while on the kidney transplant waitlist (Allen, 2023). So, Jeff and his family took matters into their own hands. They put up billboards, reached out to celebrities, designed a personalized website, and took to social media in the hope that one person would step forward to donate a kidney.

Proponents of xenotransplantation envision a future in which Jeff would not have to go to such drastic lengths to procure a kidney. Rather than wait an uncertain amount of time for a kidney that may never come, and have to take matters into his own hands, Jeff would have been able to schedule his surgery at a time when it was convenient for him. He would have received the transplant quickly if a pig organ meeting his medical indications were available or, at most, he would have had to wait only six to eight months for a kidney to be grown in a genetically engineered pig. Things might even get so good that he would not even need to go on dialysis. What would be true of Jeff would be true of others in need of a solid organ transplant.

In order for this future to become a reality, sentient nonhuman animals (animals, hereafter) must be raised and then killed, both to bring the technology to clinical trials and then for ongoing clinical therapy. At every stage of this process, animals will be confined, manipulated, and killed. Preclinical research assesses the impact of various genetic alterations in genetically engineered pig organs by either excising and assaying the organ

DOI: 10.4324/9781003472223-3

or transplanting the organ into nonhuman primates, such as baboons. This research also requires the confinement, manipulation, and early death of animals. Xenotransplantation essentially involves animals in pain and death.

It should not be surprising that our investigation into the ethics of xenotransplantation begins with an examination of the ethical defensibility of using animals in xenotransplantation. To focus our investigation, we limit ourselves to the ethics of animal research and specifically research on genetically engineered pigs. Our reason for this limitation is both practical (e.g., page considerations) and theoretical: if xenotransplant research is unethical, then so is clinical xenotransplantation; if xenotransplant research is ethical, then it would seem that clinical xenotransplantation would also be ethical. We present readers with three positions on the moral status of animals. Proponents of the view that animals have moral status akin to human beings argue that xenotransplant research is ethically indefensible for many of the same reasons that raising and killing human beings for their organs is ethically indefensible. Proponents of the view that animals lack inherent moral status argue that xenotransplant research is ethically defensible so long as it promises a favorable risk-benefit outcome for human beings. In between these two polarizing positions are the proponents of the view that animals have some degree of moral status but not as much as human beings. On this view, animal research can be ethically defensible so long as it meets certain criteria. The bulk of this chapter examines whether xenotransplant research involving animals can satisfy accepted criteria for ethical animal research.

Animal moral status

To evaluate the moral defensibility of genetically altering, confining, and killing pigs for the sake of organ harvesting, we need to examine the moral status of pigs and other sentient nonhuman animals involved in xenotransplant research. To ask whether sentient animals, who are capable of subjectively feeling pleasure and pain and who have an interest in continued life, have moral status is to ask whether they have a status that generates constraints on our behavior toward them for their own sake which would exclude viewing them merely as a resource to be instrumentalized for our sake. If sentient animals have moral status, then their interests should be considered in ethical deliberation: human beings cannot do whatever they want to them as if their interests counted for nothing. Of course, it remains to be seen how much animal interests should count in ethical deliberation. Do their interests carry the same weight as human interests? If not, how do we weigh their interests *vis-à-vis* our own?

Strong position

According to what we call a "strong" position, animals have the same or robustly similar moral status as human beings: animal interests and life are of equal or sufficiently similar weight as human interests and life. Gary Francione, a prominent defender of the animal rights position, argues that sentience, understood to include subjective awareness and ability to feel pleasure and pain (conscious experience), grounds moral considerability: "sentience provides all we need to justify personhood for non-humans" (2020, p. 8). Although some creatures have greater cognitive capacities than others, allowing for diverse phenomenal experiences and interests, Francione argues that insofar as a creature is sentient it has equal moral standing to all other sentient creatures. This entails, for Francione, that all sentient creatures enjoy the same rights, including the right to life and the right to not be harmed.

Tom Regan (2004), another prominent defender of animal rights, appeals to a "subject-of-life" criterion for moral consideration: if an individual has a subject-of-life, that individual enjoys inherent value and has it equally with all other individuals that have a subject-of-life. What amounts to being a subject-of-life? Regan explains,

> [I]ndividuals are subjects-of-a-life if they have beliefs and desires; perception, memory, and a sense of the future, including their own future; an emotional life together with feelings of pleasure and pain; preference- and welfare-interests; the ability to initiate action in pursuit of their desires and goals; a psychophysical identity over time; and an individual welfare in the sense that their experiential life fares well or ill for them. (2004, p. 243)

Inherent value, being grounded in being a subject-of-life, grounds the possession of rights. Subjects-of-life have the same rights, including the right to not be used as a mere means for the benefit of others. Since what happens to other mammals matters to them as individual subjects, "other mammals are our equals. . . . Without a doubt, these animals have rights, if humans do" (2004, p. 117).[1]

Some proponents of the strong view argue that to afford animals equal weight in moral deliberations is to hold that animals should never be used in research: "we cannot," Gary Francione explains, "morally justify using nonhuman animals in research" (2008, p. 177). Apart from a lack of necessity for much animal research, he argues that humans and animals alike "have an interest in not being used in biomedical experiments" and that we afford humans "a right not to be used as non-consenting subjects in such experiments" (2008, p. 177).

Other "strong" view theorists, by contrast, argue that taking animal moral status seriously is compatible with research on animals. Alasdair Cochrane writes that, on his interest-based approach to animal rights, "animals cannot be said to possess a general right never to be used in any form of experimentation," and if the research promises to benefit the animal, then the animal has an interest in participating in the research (2012, p. 78). An impetus for this more modest position is that we allow research on human subjects, including infants and cognitively impaired individuals, and if animals are to be considered in moral deliberations as equal or robustly similar to human persons, including infants and cognitively impaired individuals, and some research is permissible on the latter (albeit with certain qualifications), then some research is permissible on the former.

While the debate about the ethical defensibility of some animal research within a "strong" animal rights position is ongoing, we take it that xenotransplant research is ethically impermissible for all defenders of a strong animal rights position. Just as it would be wrong for researchers to genetically alter, raise in confinement, and then kill a human being for the sake of saving another human being, so it is wrong for them to do the same to pigs. No matter how well animals are treated, and no matter how much their death can help human beings, they are being sacrificed for the sake of harvesting their organs, and this is morally wrong.

Weak position

Proponents of what we refer to as a "weak" position on the issue of animal moral status take the opposite position: they argue that nonhuman animals have no moral considerability in themselves; instead, animals are morally considerable insofar as human beings relate to them. According to some proponents of this view, animals lack moral status because they are not members of a moral community (Cohen, 2017; Oderberg, 2000; Hsiao, 2017). Morality is fundamentally about performing a certain type of *action*—namely, action in pursuit of what is good and in avoidance of what is evil. Action of this kind requires *knowledge* of what is good and what is evil, for one cannot act to pursue good and avoid evil if one does not know what good and evil are. Since morality is about action in pursuit of the good, it follows that membership in the moral community requires being capable of (a) knowing what is good and evil and (b) acting for the sake of that knowledge. Whether the moral life consists of fulfilling duties, maximizing utility, or developing certain character dispositions, morality as such is something that can only be articulated, understood, and practiced by *rational beings*. Accordingly, a being has *moral status* only if it possesses a rational nature.

Another route to a weak position would be contractarianism, according to which the moral worth of an action is determined by whether it conforms to rules that citizens would choose under certain idealized circumstances to govern their behavior. A test for identifying these rules is to imagine a veil of ignorance (Rawls, 1999). Individuals behind the veil do not know any particular details about themselves—religion, age, view of animals, race, sex, nationality, etc.—but they do know basic human needs, motivations, and desires. Each individual's goal is to identify social rules that will benefit them. Because they do not know personal details about themselves, they will create rules that protect all people and shy away from creating rules that protect only some people. It does not make sense for a person to create a rule that favors one group of people when that person does not know if they are a member of said group. The set of rules agreed to by rational individuals behind the veil of ignorance constitutes moral rules.

Contractarianism has a difficult time affording animals protections (Carruthers, 1992, 2011). Behind the veil of ignorance, contractors are going to pass rules that protect themselves, *qua* human beings. They know they are human so they are going to pass moral rules governing human interaction and behavior. Animals lack rationality, which is to say, animals lack the ability to deliberate over proposed normative rules and their consequences. As such, animals fall outside of the moral community and thereby will likely lack protections and moral status. After all, contractors do not know if they will be in a position that benefits from animal research, and they do not know their previously held beliefs about the moral status of animals. Recall, all personal beliefs about religion, politics, and animals are unknown to the contractors behind the veil of ignorance. A rational, self-interested person who is ignorant of her status in the world and antecedent beliefs about animals is not going to prohibit wholesale human uses of animals because she doesn't know if she lives a life that benefits from animal use.

However they arrive at the position, proponents of a weak position of animal moral status would grant that there is nothing intrinsically wrong with xenotransplant research involving animals: confining, manipulating, and killing genetically engineered pigs for their organs does not violate moral rights or neglect any morally salient animal-centered considerations. When evaluating the ethics of xenotransplantation, what matters is human beings and their well-being. If it is wrong to confine pigs, this is because of how such confinement affects human beings, say, by making humans inclined to be cruel to other humans. This is not to say, of course, that a proponent of a weak position will be in favor of xenotransplantation; they may oppose it for any number of non-animal-related reasons, such as zoonotic risk or social justice considerations, but their opposition would be grounded in non-animal-centered reasons.

Modest position

Standing between the strong and weak positions is what we refer to as a "modest" position in which animals are morally considerable but not as much as human beings. Defenders of a modest position appeal to the moral significance of pain and suffering—pain and suffering are bad because of how they feel, and all sentient creatures want to avoid these and experience pleasure. Sentient animals, like humans, have a welfare status (i.e., a description of how well they are getting along in the world), and an interest in a life free from suffering and coercion; other moral agents are to respect these interests and cannot violate them for no good reason. As such, human beings cannot treat animals however they want with no regard for their interests and well-being. There are moral limits on how we ought to treat them. However, defenders of a modest position grant that human beings have greater weight in moral deliberations than animals. For instance, if you have a choice between saving a drowning child or a drowning kitten, you should save the child—preference should be given to the child, even though both the child and the kitten have an interest in being rescued. Why exactly this is the case is subject to dispute, but a common view is that humans have greater moral considerability on account of their more robust cognitive capacities (e.g., rationality, language, or sociality) or having been made in the image of God, or having an autonomous will.

Whether xenotransplantation is ethically defensible on a modest view is an open question. To determine an answer requires an assessment of the ethics of animal research. Could the confinement, manipulation, and early death of an animal be ethically defensible? How do we compare potential human benefit with animal suffering and death? For purposes of this book, does xenotransplant research satisfy the conditions of ethical animal research? We turn to such questions in what follows.

Principles of ethical animal research

While there are competing accounts of what makes for permissible animal research, that is, how to weigh human versus animal interests, we focus on the account advanced by Tom Beauchamp and David DeGrazia in their influential 2020 book, *Principles of Animal Research Ethics*. Since their account sets a high bar for the permissibility of animal research, examining xenotransplantation in light of it should be illuminating. Beauchamp and DeGrazia offer three principles of social benefit and three principles regarding animal welfare. After explaining each principle, we set out what critics and defenders of xenotransplantation may say in light of it (Bobier et al., 2023).

The principle of no alternative method

According to the first principle, animal research is ethical only if there is no ethically acceptable alternative: "the use of animal subjects," Beauchamp and DeGrazia state, "is justified only if the knowledge and social benefits pursued in a research study are not ethically obtainable through alternative research methods that do not use live animals" (2020, p. 7). This principle is compelling: if research A involves animals in pain, suffering, and death while research B does not, then, all things considered, it is preferable to pursue research B. "If, for example," they write, "methods that use human tissues, cell cultures, or computer models are sufficient, sentient animals must not be used" (2020, p. 7). Applied to xenotransplantation, the issue is whether there are alternatives that avoid using animals and yet promise to be effective at contributing to the organ supply.

Critics of xenotransplant research argue that this principle is not satisfied, for there are other possible solutions to the organ supply shortage that do not require the use of animals. At the most basic level, there is a need to increase access to preventative medicine and promote educational awareness of the importance of lifestyle changes to improve public health and drive down the demand for organs (Menache, 1999). This would not eliminate the need for organ transplantations, of course, and so this would have to be coupled with other proposed solutions, which vary in their feasibility and acceptability. These possibly include building more Intensive Care Units, training more surgeons, expanding donation criteria, furthering research on lab-grown organs, and creating a market for kidney transplants. Many of these options were discussed in Chapter 1. What matters presently is that, although these possible solutions involve their own ethical issues, they do not involve the general problems raised by animal welfare and death that are integral to xenotransplantation (Johnson, 2022a).

Proponents of xenotransplant research agree that there are other possible solutions that do not involve confining, manipulating, and killing animals for organ harvesting. They also agree that, all things considered, if one solution does not involve animal harm and the other does, then the harmless-to-animals solution should be pursued. But why the proponent of xenotransplantation does not think this line of research should cease is because she remains unconvinced that harmless-to-animals solutions are adequate individually or collectively to meet the organ deficit. Proponents argue that, ultimately, it is likely that the organ shortage can only be addressed through a multifaceted approach, whereby xenotransplantation will be one of the approaches utilized.

How might xenotransplant proponents argue for this claim? The answer depends on whether xenotransplantation can help address the organ deficit soon and on whether there is evidence that alternative solutions are not enough. Although some of these are already covered in Chapter 1, it is worth reiterating:

- There is evidence to suggest that the alternative policies may not be enough or are unlikely to be implemented. Studies of the implementation of policies aimed to promote organ donation in Singapore, Canada, and Chile found little to no improvement in donation rates (Gill et al., 2008; Lee et al., 2009; Wynn and Alexander, 2011). In a comparison study between countries that adopt an opt-in versus an opt-out organ donation policy, Arshad et al. report, "our data demonstrate no significant difference in deceased donation or solid organ transplantation activity between opt-out versus opt-in countries" (2019, p. 1453). Other policy solutions such as organ markets are unlikely to be implemented anytime soon.
- Scientists are much further along in creating transplantable solid organs in genetically modified pigs than they are in creating lab-grown solid organs for transplantation. While kidney xenotransplant clinical trials are likely to begin soon, the beginning of trials for functional lab-grown organs—for example, 3D printed solid organs—is still far off and practically out of sight.

Xenotransplant proponents are in favor of pursuing alternative methods to addressing the organ deficit, but do not think this should come at the expense of xenotransplant research: if, in the future, researchers are able to grow functional solid organs in a lab, then this would, all things considered, be preferable to xenotransplantation. However, in the meantime, xenotransplant research is justified because it promises to help save human lives.

The principle of expected net benefit

According to the second principle, the social benefits that are expected to result from research involving animals should outweigh the expected risks to human beings. The focus of this principle is human beings, as Beauchamp and DeGrazia explain:

> If investigators do not present a credible case that the prospect of benefit exceeds expected costs and risks, inadequate grounds exist to support a claim that the research is justified and worth pursuing even from a standpoint that considers only human interests.
>
> (2020, p. 9)

To assess this principle, researchers should take into consideration (a) the magnitude of the benefits if the intended outcome were achieved *vis-à-vis*

(b) the likelihood of achieving the intended benefits: how likely is the research to result in a social benefit and how significant would the benefit be? To assess the likelihood of success, Beauchamp and DeGrazia posit that researchers should compare "the frequency with which comparable research has previously yielded or failed to yield benefits" to the relative successes (2020, p. 8). This should then be compared to (c) the anticipated costs and risks for human beings.

Proponents and critics of xenotransplantation can agree that the magnitude of the benefit is high: it promises to provide a new supply of safe, effective, and scalable organs. Their disagreement will be over (a) how likely it is to become clinically viable and (b) how to weigh this benefit with the risks associated with it. Undoubtedly, there are scientific hurdles to overcome in order for xenotransplant research to progress to and beyond clinical phases. Critics have at least three concerns:

- *Animal studies*: It is difficult to extrapolate from preclinical xenotransplant research in nonhuman animals to human subjects due to the physiological differences between humans and nonhuman animals (Reichart et al., 2022). Moreover, animal-to-animal xenotransplant research assesses pig organs that have been engineered to evade human immune responses, not primate immune responses, thereby limiting the value of this line of research. Such limitations have practical significance, as Montgomery et al. (2024, p. 522) note, nonhuman primate data showed "100% primate survival for over 8 months," and yet, "the first 2 recipients of pig hearts died unexpectedly within 2 months."
- *Human-decedent studies:* Researchers have recently begun studying xenotransplantation on human bodies of former persons who have been declared dead by neurologic criteria. The validity of such research is contested, for human bodies sustained artificially may not be indicative (Cooper and Kobayashi, 2024), and the research is often terminated early to allow the family of the deceased to mourn (Adams et al., 2023). Extrapolating meaningful data is thereby difficult.
- *Human results:* Recent xenograft recipients have not lived long and were not subject to the rigors of a standard clinical research. The 2022 cardiac xenotransplant into a severely ill patient, David Bennett Sr., lasted only 60 days before he died, while the second cardiac xenograft recipient, Lawrence Faucette, died just shy of six weeks posttransplant (Montgomery et al., 2024). The first kidney xenograft recipient, Richard "Rick" Slayman, died seven weeks posttransplant (Hassan, 2024). That these individuals died shortly after is troubling, as is the fact that the condition under which the xenotransplants occurred—the Food and Drug Administration's compassionate use program—does not require the rigor of a clinical trial.

Proponents of xenotransplant research may respond that research is progressing at an accelerating rate and that, even if the research thus far has not been as promising as critics would like, there is a growing body of data and phaseless kidney xenotransplant clinical trials have been approved. So while there are hurdles to overcome, xenotransplant proponents posit that there is no reason to doubt that it will be clinically viable.

What about the anticipated risks and costs to society? A primary risk is the potential for xenozoonotic diseases being transferred from genetically modified pigs to human xenograft recipients and ensuing potential for an epidemic or pandemic event. Even if xenotransplantation becomes clinically viable, there is the ever-present concern that transplanted pig organs will result in a pandemic of unknown but potentially serious risk. In most ordinary cases of research, the research poses a threat to an individual person; in xenotransplantation, however, the whole world is, plausibly, at risk. This is a serious concern that we will devote Chapter 3 to.

The principle of sufficient value to justify harm

According to the third principle, researchers need to establish that the harm to animals is offset by sufficiently valuable benefits to human beings: "research animals should not be harmed for inconsequential reasons" (2020, p. 10). It is unclear what amounts to an inconsequential reason, which is why Beauchamp and DeGrazia leave it to researchers and review committees to "judge how much prospective net benefit to society counts as sufficiently valuable to justify anticipated harms to the research participants" (2020, p. 9). The underlying idea is plausible, namely, trivial benefit does not justify significant animal harm, but guidance to aid researchers and review committees in their assessment of various proposals is nevertheless lacking.

The suffering and death experienced by genetically engineered pigs should not be taken lightly. Bernard Rollins observes that, although their welfare will be better than most other pigs, their living situation will "be equally deficient in accommodating the animals' biological and psychological natures" (2020, p. 4). Johnson is explicit as to why this may be:

> These pigs are genetically modified and cloned, and must be bred and housed using infection-control measures like artificial insemination, embryo transfer, Caesarian births, and isolation in sterile environments without contact with other animals, preventing the expression of their natural behaviors. Their use would require frequent blood and tissue sampling, which in pigs requires restraint, including drug-induced restraint.
>
> (2022b, np)

The confinement, manipulation, and early death of pigs matter morally. It is admittedly nebulous how to weigh the moral status of animals with the benefit to humans, but in light of the promise of xenotransplantation to help mitigate the organ crisis, it is plausible that the expected benefit to human beings does in fact justify the harm befalling pigs. If any animal research is morally permissible, it is permissible on account of saving human lives, which is exactly the hope of xenotransplant research. Of course, this assessment hinges on there being no other viable therapeutic solution, which is subject to dispute.

The principle of no unnecessary harm

The fourth principle posits that animal research subjects should not be unnecessarily harmed. Their suffering and death must be integral or inherent in the research, as Beauchamp and DeGrazia write, "Animals should not be caused pain, distress, suffering, or the like in experimental procedures unless the harm is unavoidable because of morally justified scientific objectives that satisfy the principles of social benefit" (2020, p. 13). The harm involved should be minimized as much as possible in accordance with sound scientific principles and aims. This principle is grounded in the strong moral intuition that causing pain and suffering without good reason is wrong and that there should be no needless harm resulting from negligence.

This principle is well-taken by proponents and critics of xenotransplantation alike—all agree that unnecessary harm should be eliminated. Medical organizations and governments call for high welfare standards in xenotransplant research (e.g., FDA, 2016). The challenge of assessing this principle is determining what counts as "unnecessary" harm, a challenge that besets ordinary clinical care, even under the best of conditions. The nature of the research warrants blood sampling, a sterile environment, and other lack of niceties, which is not to be confused with an environment devoid of toys or conspecifics. Not testing for infection threatens the well-being of human beings and pigs alike. In general, if there are ways to promote genetically engineered pig welfare without compromising safety, then xenotransplant proponents ought to be in favor.

Xenotransplant proponents can argue that genetically engineered pigs will be treated well independently of legal regulations: their welfare helps to ensure better results in advancing the primary goal of supplying healthy organs for transplant. Arguably, it is in the interest of biotechnology companies—due, at a minimum, to public relations—to permit genetically engineered pigs to have the best possible lives within the confines of what is possible, reasonable, and safe. Moreover, unhealthy, stressed pigs may compromise organ quality, and in order to maximize the health of the transplantable organs and minimize zoonotic concerns, genetically engineered pigs will not be housed in cramped, miserable cages, nor will they live in their own waste as is often the case on industrialized farms. To reduce psychological and physiological stress that can be transmitted to organs, genetically engineered pigs will have stimulation, including toys and conspecifics. They will not be slaughtered in massive slaughtering factories that increase stress and pain; rather, in order to ensure the safe retrieval of organs, genetically engineered pigs will be as painlessly sedated as possible. Since, according to the xenotransplant proponent, there is a compelling reason inherent in xenotransplant research to treat genetically engineered pigs as well as possible and to minimize pain, the burden of proof is on the xenotransplant critic to show that there is in fact unnecessary harm occurring.

The principle of basic needs

The fifth principle is that the pigs' basic needs should be met unless there are unavoidable and morally justified limitations that satisfy the three social principles above. The underlying motivation, Beauchamp and DeGrazia explain, is that satisfaction of "basic needs is necessary for animals to have a minimally decent quality of life" (2020, p. 14). This means that genetically engineered pigs are not unnecessarily deprived of basic goods, such as water, food, socialization, mobility, adequate space, freedom from injury, and the like. While death is required by the research, as are other kinds of manipulation and control, other basic goods should not be needlessly deprived.

There is little reason to believe that genetically engineered pigs' basic needs are not being met and it would be difficult to argue so. "There is a widespread agreement," Marı´a Jorqui-Azofra explains, "that high standards of animal welfare" are of high importance. Cozzi et al. found that facilities "for rearing specific pathogen-free (SPF) . . . pigs follow high welfare and safety standards" (2020, p. 321), and Martelli et al.'s study on genetically engineered pigs found no significant differences on various welfare metrics compared to nongenetically engineered pigs, suggesting to them that there is "no significant undesirable effect" as a result of genetic engineering (2014, p. 815). Researchers have reason to maintain high welfare standards and no reason not to: stress on the organ, health of the pig, public perception, legal compliance, and so on.

The principle of upper limits to harm

The sixth and final principle is that animals should not be "expected to endure severe, long-lasting suffering" (2020, p. 16). There is an upper limit to the amount of suffering a morally considerable creature can suffer. Animal subjects should not severely suffer for an extended period of time, no matter the outcome. The idea is that animals' lives are, on the whole, positive or of sufficient quality so as to *not* render killing them an act of a kindness: researchers should not subject animals to un-ameliorable and lasting pain.

We assume that proponents and critics alike agree with this condition. The issue is whether genetically engineered pigs will suffer for a long duration and whether this suffering is ameliorable through appropriate anesthesia or analgesics. While few are in a position to know the extent and duration of genetically engineered pig suffering, we do not think, based on evidence already cited, that this condition is unmet in xenotransplant research. At the very least, critics would need to marshal compelling evidence of long-lasting and unmitigable suffering in these pigs.

The case for further genetic engineering

Considerations of the suffering that genetically engineered pigs may experience have led some scholars to suggest a solution: researchers should further engineer pigs to reduce their capacity for suffering (Rodger et al., 2024). As with other facets of genetic engineering, this is not a far-fetched possibility, for it has been demonstrated that researchers can modify affective pain or felt pain awareness but not nociception, and there is a growing literature in defense of welfare-based arguments for engineering out animals' capacity to experience pain, specifically animals used in industrialized animal agriculture and animal research more broadly. The idea is seductively simple: if we engineer pigs to lack the enzymes in the brain that are responsible for affective pain experiences, then they would not experience subjective, phenomenal pain, though they may still experience pleasure and display pain-related behavior. This would result in genetically engineered pigs living brief lives of some pleasure and little to no phenomenal pain, and thus, there will not be pressing animal welfare concerns. Moreover, the high degree of existing genetic manipulation required to engineer pigs for xenotransplantation means that, were it scientifically possible to achieve this outcome, there would be ample opportunity to do so.

There are two possible routes researchers could take. First, the pigs may be rendered insentient in such a way that they can grow like normal pigs but have no higher level brain function, that is, animal microcephalic lumps. This would eliminate all pain and suffering. Or, second, the engineering can be more piecemeal, for example, by moderating the corticotropin-releasing hormone (CRH), which affects cortisol production and behavioral displays of stress. While there may be concerns that piecemeal engineering may promote *more* suffering than otherwise, this seems to be an empirical question, one that requires researchers to breed such animals and observe their welfare. Importantly, further genetic engineering—along with well-designed designated pathogen-free facilities and evidence-based animal husbandry practices—can help to address some of the animal welfare concerns with xenotransplant research that have been raised, namely, the concerns that this research is inherently unpleasant and painful for pigs (Entwistle et al., 2022; Johnson, 2022b).

How might a critic respond? Objections to genetic disenhancement such as that under consideration tend to be blanket objections. Some argue that it violates the integrity of the animal (Bovenkerk et al., 2002), or that it is part of a broader system of oppression (Henschke, 2012). Murphy and Kabasenche (2018), for instance, argue that if society valued the capacities of animals, we would try to improve the experience of animals, rather than removing their capacity to experience. Such objections entail the indefensibility of xenotransplantation altogether and not further genetic engineering out of experienced pain in particular. In other words, if genetic engineering of pain receptors is wrong, then so is xenotransplantation.

In sum, there are two positions a critic could take. On the one hand, the critic can oppose xenotransplant research and be in favor of nonanimal-harming alternatives: this critic believes that such research is wrong for animal-related reasons. On the other hand, the critic can recognize that xenotransplant research is ongoing and is likely going to continue. The research train has already left the station, so to speak. As such, the critic can agree that it is better for pigs that their ability to experience pain is eliminated or otherwise minimized than not, even if the whole institution is ethically problematic. This is compatible with continuing to argue against the ethical defensibility of xenotransplant research; however, this can be uncomfortable and the critic may rest content with arguing against xenotransplantation.

Conclusion

Whether the case for or against xenotransplant animal research is successful, we leave up to the reader to decide. A decision should not be made on the basis of this chapter alone, however, for the ethical defensibility of xenotransplantation relies on one's assessment of and response to the risk posed by potential xenozoonotic infection. This is the focus of the next chapter: how should we respond to the potential risk of xenozoonotic infection? In closing, we want to reiterate that *if xenotransplant animal research is ethically defensible, then there is no animal-centered concern about clinical xenotransplantation regarding animal use*. There might be other ethical concerns with it, concerns we discuss in the following chapters; our point is that, if confining, manipulating, and killing an animal for xenotransplant research is ethically defensible, then confining, manipulating, and killing an animal for xenotransplant clinical practice is ethically defensible. Indeed, an ethical defense of xenotransplant research requires more rigor since research may not directly benefit a person whereas clinical xenotransplantation does directly benefit a person: if such research, which may not directly a person, is justified, then the ensuing clinical practice, which does benefit a person, is justified.[2]

Notes

1 There are other avenues to a strong position apart from an animal rights framework. Peter Singer, a well-known utilitarian, rejects the existence of moral rights but agrees that "the basic principle of equality" extends to sentient nonhuman animals (2009, p. 2): sentient animals have an interest in not suffering like we do, and "the principle of equality requires that" all suffering "be counted equally" (2009, p. 8). Care ethicists, likewise, ground moral considerability in the kinds of relations others, including sentient animals, can have with us. Being in a moral relation with another involves honest attempts to appreciate and respond to another's situation, including their needs, pains, and desires (Gruen, 2015). This involves cultivating "empathetic imagination"—empathizing with another's situation and imagining oneself in it—so that we can foster and deepen our relationships (Gruen, 2015). Care ethics has been used to show that we have strong moral obligations to animals—we have moral obligations to animals because we care about them, and this includes not using or otherwise radically reimagining scientific practice. We set these and other similar positions to the side and focus on animal rights theorists.

2 We would like to thank Daniel Rodger for his helpful feedback on this chapter.

References

Adams, A., Cendales, L. C., Cooper, D. K., Cozzi, E., Gill, J., Judd, E., Katz, E., Kirk, A. D., Fishman, J. A., Reese, P. P. and Wall, A. 2023. American society of transplant surgeons-American society of transplantation report of FDA meeting on regulatory expectations for xenotransplantation products. *American Journal of Transplantation*, 23(9), 1290–1299.

Allen, J. 2023, May 17. 'Please save my life': Colorado man on transplant waiting list turns to billboard asking for a kidney. *ABC Denver 7*. https://www.denver7.com/news/contact-denver7/man-marketing-himself-for-kidney-through-brighton-billboard-car-decals (Accessed: 20 March 2024).

Arshad, A., Anderson, B. and Sharif, A. 2019. Comparison of organ donation and transplantation rates between opt-out and opt-in systems. *Kidney International*, 95(6), 1453–1460.

Beauchamp, T. L. and DeGrazia, D. 2020. *Principles of Animal Research Ethics*. Oxford University Press.

Bobier, C., Rodger, D., Hurst, D. J. and Omelianchuk, A. 2023. In defense of xenotransplantation research: Because of, not in spite of, animal welfare concerns. *Xenotransplantation*, 30(1), e12791.

Bovenkerk, B., Brom, F. W. and Van Den Bergh, B. J. 2002. Brave new birds: The use of 'animal integrity' in animal ethics. *Hastings Center Report*, 32(1), 16–22.

Carruthers, P. 1992. *The Animals Issue: Moral Theory in Practice.* Columbia University Press.

Carruthers, P. 2011. Against the moral standing of animals. *Questions of Life and Death: Readings in Practical Ethics*, 274–284.

Cochrane, A. 2012. *Animal Rights Without Liberation.* Columbia University Press.

Cohen, C. 2017. The case for the use of animals in biomedical research. In *Research Ethics.* Routledge, 459–464.

Cooper, D. K. and Kobayashi, T. 2024. Xenotransplantation experiments in brain-dead human subjects: A critical appraisal. *American Journal of Transplantation*, 24(4), 520–525.

Entwistle, J. W., Sade, R. M. and Drake, D. H. 2022. Clinical xenotransplantation seems close: Ethical issues persist. *Artificial Organs*, 46(6), 987–994.

Food and Drug Administration. 2016. *Source Animal, Product, Preclinical, and Clinical Issues Concerning the Use of Xenotransplantation Products in Humans: Guidance for Industry.* https://www.fda.gov/files/vaccines,%20blood%20&%20biologics/published/Source-Animal-Product-Preclinical-and-Clinical-Issues-Concerning-the-Use-of-Xenotransplantation-Products-in-Humans-Guidance-for-Industry.pdf.

Francione, G. L. 2008. *Animals as Persons: Essays on the Abolition of Animal Exploitation.* Columbia University Press.

Francione, G. L. 2020. *Why Veganism Matters: The Moral Value of Animals.* Columbia University Press.

Gill, J. S., Klarenbach, S., Cole, E. and Shemie, S. D. 2008. Deceased organ donation in Canada: An opportunity to heal a fractured system. *American Journal of Transplantation*, 8(8), 1580–1587.

Gruen, L. 2015. *Entangled Empathy: An Alternative Ethic for Our Relationships with Animals.* Lantern Books.

Hassan, J. 2024. Man who received pig kidney transplant dies weeks after milestone surgery. *The Washington Post*, May 12. https://www.washingtonpost.com/health/2024/05/12/pig-kidney-transplant-dies-richard-slayman/

Henschke, A. 2012. Making sense of animal disenhancement. *NanoEthics*, 6, 55–64.

Hsiao, T. 2017. Industrial farming is not cruel to animals. *Journal of Agricultural and Environmental Ethics*, 30, 37–54.

Johnson, L. S. M. 2022a. Existing ethical tensions in xenotransplantation. *Cambridge Quarterly of Healthcare Ethics*, 31(3), 355–367.

Johnson, L. S. M. 2022b. Xenotransplantation: Three areas of concern. *Hastings Bioethics Forum.* https://www.thehastingscenter.org/xenotransplantation-three-areas-of-concern/

Jorqui-Azofra, M. 2020. Regulation of clinical xenotransplantation: A reappraisal of the legal, ethical, and social aspects involved. *Xenotransplantation: Methods and Protocols*, 315–358.

Lee, V. T., Yip, C. C., Ganpathi, I. S., Chang, S., Mak, K. S., Prabhakaran, K. and Madhavan, K. 2009. Expanding the donor pool for liver transplantation in the setting of an "opt-out" scheme: 3 years after new legislation. *Annals of the Academy of Medicine of Singapore*, 38(4), 315–317.

Martelli, G., Sardi, L., Stancampiano, L., Govoni, N., Zannoni, A., Nannoni, E., Forni, M. and Bacci, M. L. 2014. A study on some welfare-related parameters of hDAF transgenic pigs when compared with their conventional close relatives. *Animal*, 8(5), 810–816.

Menache, A. 1999. Stop now before it's too late. *Bulletin of the World Health Organization*, 77(1), 76–78.

Montgomery, R. A., Griesemer, A. D., Segev, D. L. and Sommer, P. 2024. The decedent model: A new paradigm for de-risking high stakes clinical trials like xenotransplantation. *American Journal of Transplantation*, 24(4), 526–532.

Murphy, K. N. and Kabasenche, W. P. 2018. Animal disenhancement in moral context. *NanoEthics*, 12, 225–236.

Oderberg, D. S. 2000. *Applied Ethics: A Non-Consequentialist Approach*. Wiley Blackwell.

Rawls, J. 1999. *A Theory of Justice*. 2nd edition. Belknap Press.

Regan, T. 2004. *The Case for Animal Rights: Updated with a New Preface*. University of California Press.

Reichart, B., Cooper, D. K., Längin, M., Tönjes, R. R., Pierson III, R. N. and Wolf, E. 2022. Cardiac xenotransplantation: From concept to clinic. *Cardiovascular Research*, 118(18), 3499–3516.

Rodger, D., Hurst, D. J., Bobier, C. A. and Symons, X. 2024. Genetic disenhancement and xenotransplantation: Diminishing pigs' capacity to experience suffering through genetic engineering. *Journal of Medical Ethics*, 50(11), 729–733.

Rollin, B. E. 2020. Ethical and societal issues occasioned by xenotransplantation. *Animals*, 10(9), 1–10.

Singer, P. 2009. *Animal Liberation: The Definitive Classic of the Animal Movement*. Harper.

Wynn, J. J. and Alexander, C. E. 2011. Increasing organ donation and transplantation: The US experience over the past decade. *Transplant International*, 24(4), 324–332.

3 What are the public health concerns?

In 1906, Charles Henry Warren, a wealthy New Yorker, commissioned a sanitary engineer by the name of George Sober to investigate the source of a typhoid fever (*Salmonella typhi*) outbreak at his residence: 6 of 11 residents were suffering from typhoid fever. At the time, there were no antibiotic treatments or immunizations for typhoid and it was fatal in 10% of cases. Initially suspecting clams as the culprits, Sober noticed an unlikely coincidence: the infected individuals all interacted with Mary Mallon, an immigrant from Ireland and a kitchen worker on Long Island. His curiosity piqued, Sober soon discovered that Mary initially had a mild form of typhoid but remained an asymptomatic carrier of it, transmitting it to those she came into contact with. In 1907, about 3,000 people in New York City were infected with typhoid and all fingers pointed at Mary as the likely culprit. Despite being asymptomatic, but wanting to protect the public from typhoid fever, public health officials forcibly quarantined Mary on North Brother Island for well over 20 years.

The case of Mary Mallon is illustrative of fundamental questions in public health ethics (Bourdain, 2001). At what point does the autonomy of the individual become overridable on the basis of the well-being of others? When, if at all, should personal information be disclosed to third parties without consent? When, if at all, does public risk override personal benefit? Mary was asymptomatic and yet she was forced to quarantine against her will for over 20 years to protect others, not herself.

When discussing the ethics of using genetically engineered pigs for organ transplantation in the previous chapter, we noted that transplanting genetically modified organs poses a potential public health risk, thereby complicating the potential risk-benefit analysis of this research. In this chapter, we examine this issue in detail. We begin by explaining the potential risk of a pandemic of unknown severity that xenotransplantation poses. We then engage three responses to the public health risk. Some argue that xenotransplant research should cease for precautionary reasons; some argue that it

DOI: 10.4324/9781003472223-4

requires the collective consent of everyone, not merely the recipient; and some argue that xenotransplant recipients should waive their right to withdraw from research.

Xenotransplantation and the risk of zoonosis

Transplanting organs from genetically engineered pigs poses the potentially serious, or even catastrophic, risk of xenozoonosis—the transmission of a microorganism from the host animal to humans via the xenograft that is pathogenic in humans. This is not merely a theoretical concern, for the vast majority of emerging infectious diseases are zoonotic, and diseases such as HIV, Rabies, West Nile Virus, Swine Flu, and Ebola have resulted in significant loss of human life over the years; zoonotic diseases are estimated to account for more than 2.5 million deaths every year (Salyer et al., 2017). While the infection risk posed by xenotransplantation is considered to be low, genetically engineered pigs may still present an infectious disease risk due to various swine-derived pathogens. Pigs are carriers of porcine endogenous retroviruses (PERVs) and are susceptible to porcine cytomegalovirus (PCMV), both of which pose risks to human health. PCMV is a common herpesvirus and, as far as is known, does not infect human cells but can infect, and become pathogenic, in porcine cells. PERVs, by contrast, have become integrated into the pig genome. As Łopata and colleagues explain, due to their evolutionary role in protecting porcine health, "PERVs constitute an integral part of the porcine genome and are present in various proportions depending on pig breed, tissue type, and retrovirus subtype" (2018, p. 1). There are three types of PERVs, two of which are polytropic viruses that infect cells of different species and are present in all pigs (Denner, 2016, 2021). That PERVs are in the porcine genome means that they cannot be mitigated through rearing practices or embryo selection, and while they are generally harmless to pigs, they may be fatal in other species (Denner, 2021).

There is thus the ever-present concern that a xenograft recipient may become infected with a swine-derived pathogen, including but not limited to PERVs, that could go undetected (Nellore et al., 2022). A PERV could, Sade and Mukherjee explain, "infect human cells, mutate, and cause cancer, or combine with other viruses to cause novel infectious diseases" (2022, p. 713). A xenozoonotic disease could express itself soon after transplantation, as seems to have happened in the severely ill recipient of a genetically engineered porcine heart, David Bennett Sr., who died two months after the transplant; it is believed, but still unconfirmed, that a previously undetected xenozoonosis—PCMV—was partly to blame (Fischer and Schnieke, 2022). Because zoonoses may have long latency periods before expression, a microorganism could be transmitted but remain dormant within a recipient for a period of time before ever becoming pathogenic (Denner, 2022, 2016). The xenograft recipient is thus always at risk, even if the transplant fails and the xenograft is removed.

Xenozoonotic transmission could theoretically spread from the xenograft recipient to others, possibly leading to an epidemic or pandemic. As Johnson explains, "everyone in the world is at risk from an XTx-related [xenotransplantation-related] infection, not merely the individual xenograft recipient" (2022, p. 360); she goes on to say the "unknown and unquantifiable risks of [xenotransplantation] include the possible unleashing of zoonotic diseases that could potentially affect the entire world" (2022, p. 364). Generally, xenotransplantation researchers believe there is a low risk of xenozoonotic infection occurring in a xenograft recipient and then spreading to the broader public; however, many unknowns remain. In reality, only clinical trials can determine the true nature of this risk—the potentially global scale of the risk warrants obvious caution and whether or when they should begin rightly remains contentious and the source of ongoing debate (Rodger et al., 2023).

Precautionary argument

What does the risk of a potential pandemic of unknown severity ethically require of researchers and clinicians? One response is that it requires researchers to stop, to prevent the possibility of zoonosis from arising. In an early but important paper, Fovargue and Ost (2010) argue that clinical trials in xenotransplantation should be prohibited based on precautionary reasoning: the nonzero risk of a novel infectious disease pandemic means that permitting clinical trials would knowingly expose the population to an unknown and unprecedented risk that could not be outweighed by the benefits to an individual recipient. Stated differently, significant potential third-party risk overrides the lifesaving possibility of xenotransplantation.

Xenotransplant proponents have a response. Precautionary reasoning is commonsensical insofar as the status quo is acceptable; there should be a presumption against putting the status quo at risk through uncertain decisions, so the burden is on innovators to show that their research will not increase the overall risk to it. The tendency, though, is just to assume the status quo is good, which it may not be, and to only focus on the potential risks of the research and none of its potential benefits while offering no guidance about what level of safety is adequate. While some precautionary measures are warranted in the face of uncertainty, uncertainty by itself, even of a worst-case scenario, does not generally justify a moratorium on potentially beneficial research. If it did, then the only acceptable forms of research would be those that pose no risk whatsoever, an impossible standard to meet. We need to continue to learn more about the likelihood of the risks involved, and at this point, there is not enough evidence to demonstrate that the risks posed by xenotransplantation are unacceptable.

It must be remembered that precautionary measures incur their own costs, in terms of both forgoing and delaying the potential benefits of an activity while simultaneously imposing their own burdens on those required to meet them. In the case against xenotransplantation, further extending the known mortality and morbidity of people awaiting organ transplant must be accepted to ensure a "no risk allowed" safety standard to avoid an unknown threat of a pandemic that could very well be preventable or mitigable. In an earlier analysis, Veatch and Ross concluded that, "at this point," the precautionary approach to xenotransplant is "overly burdensome relative to the potential benefits that such research could provide" (2015, p. 245). Since then, the potential benefits appear closer than ever to being realized.

Lastly, researchers are seeking ways to minimize, if not mitigate, zoonotic risk. For example, one study showed that early weaning of piglets eliminated PCMV altogether (Egerer et al., 2018). Not all zoonotic diseases can be eliminated through rearing techniques, because PERVs are found in the pig genome. Fortunately, research is progressing on better detection and genetic deletion of PERVs in genetically modified pigs. One team of researchers were able to gene-edit pigs resistant to classical swine fever virus (Xie et al., 2018), while another team of researchers were able to inactivate all 25 PERV copies, which were then used in somatic cell nuclear transfer to produce piglets; no reinfection was observed (Niu et al., 2017). Yet another team of researchers succeeded in inactivating 62 copies of proviruses in the pig genome, creating PERV-inactivated pigs (Yang et al., 2015). Moreover, certain antiretroviral drugs and vaccines are available and could be used to prevent PERV infection, as well as RNA interference technologies (Denner, 2021). In any event, the results from research on the transmissibility of retroviruses should function as a limiting factor for xenotransplantation. If it proves to minimize risk to acceptable levels, and it certainly appears that way, then it should go forward; if not, then not.

That said, it is important to bear in mind that many of the zoonotic concerns are still conjectural at this point in time. As Denner explains,

> no PERV transmission has been observed in clinical trials transplanting pig islet cells into diabetic humans, in preclinical trials transplanting pig cells and organs into nonhuman primates with remarkable long survival times of the transplant, and in infection experiments with several animal species.
>
> (2021, p. 1)

PERV infection has only ever been observed in vitro, not in vivo. While this is good news, it must be balanced with the report that the first genetically altered porcine heart recipient did, unexpectedly, have a PCMV infection which is associated with a reduction in survival times of porcine xenografts in nonhuman primates. Yet the value of this discovery is that we now know that more sensitive methods for screening PCMVs are needed.

Collective consent

Another response to the risk of potential pandemic of unknown severity is to require that researchers obtain the consent of all who are at risk. Given that xenotransplantation has the potential to cause significant harm to individuals the world over, it has been argued that ethical xenotransplantation requires the consent of everyone that may be affected: collective consent is required in addition to due diligence. The argument for collective consent can be spelled out as follows. In situations in which researchers can potentially benefit an individual but only by imposing potentially serious risks on others, concerns of collective harm outweigh individual benefits. When duties of nonmaleficence and duties of beneficence come into conflict, nonmaleficence wins out when nonmaleficence is of interest to the many and beneficence to the few: researchers should not place the many in danger to potentially help the few. To place many in danger to save one or a few is ethically permissible only if the many consent to being placed in danger, which is to say, they are, in the words of Jorqui-Azofra, "given an opportunity to participate actively and meaningfully in the decision about whether and under what conditions they are exposed to the risk" (2020, p. 330). To place the many in danger without their consent is to disrespect their autonomy, and this is for the same reason that it would be wrong for a researcher to place a research participant in danger without his or her consent. Consequently, as the argument goes, for xenotransplant research to be ethical, there needs to be collective consent to it: "xenotransplantation," Fovargue concludes, "necessitates some form of third-party consent" (2011, p. 183).

The conclusion that collective consent should be obtained for xenotransplant research to ethically proceed is appealing at first sight. However, there is a serious problem with the proposal that researchers must obtain the actual consent of all who are at risk, that is, everyone in the world: "consent of all those potentially involved in the harm at stake," Ravelingien notes, "is virtually impossible" (2007, p. 131). A common thought is that Kantian ethics, which serve to justify the need for consent in the first place, does not demand the impossible of us, that if we should do something, then we ought to be able to do it. Since obtaining the consent of every person in the world is not possible, how can researchers be ethically required to do this?

There are three responses to this practical concern. First, it can be maintained that xenotransplant research is unethical and, therefore, should not proceed. This conclusion follows from granting that collective consent is necessary for xenotransplant research to be ethically permissible and that it is impossible to obtain such consent. Defenders of this position may argue that alternatives to xenotransplantation that do not place the public in danger should be pursued to mitigate the organ shortage. Johnson writes that alternative solutions "to the organ shortage do not implicate the several significant ethical concerns for humans and animals, and for that reason are much preferable" to xenotransplantation (2022, p. 366). Others may argue that xenotransplant research should stop until further preclinical and genetic research establishes a much more favorable risk-benefit ratio, for example, the near total elimination of risk of zoonotic disease (Menache, 1999). This would be a high bar to meet, but if it could be met, then collective consent would not be necessary since the collective risks would be mitigated.

Not everyone thinks xenotransplant research should cease, and so, the challenge becomes identifying how to ground the possibility of collective consent. This leads to the second response: it could be argued that all that is required is the consent of authorized representatives, for example, medical organizations or state departments of health (Jorqui-Azofra, 2020). This would be easier in theory to obtain than the consent of every person in the world, but this kind of substituted consent is too weak to provide a basis for giving permission to researchers to move forward given the argument earlier. Xenotransplantation could potentially place the wider population at risk, and as such, ethical xenotransplantation requires *your* consent, not the consent of a medical organization or health department on your behalf. Imagine researchers conducting potentially risky research on a group of competent decision-makers without their consent on the grounds that a governing authority permitted it. If such research becomes known to the competent decision-makers, they have a right to complain that the research was unethical on the grounds that their autonomy was not respected. Or consider researchers conducting research on an individual, but the research places you, your family, your friends, and your community at potentially considerable risk. Given that you are at risk and you derive no benefit, you should have a say in whether the research moves forward. So, it appears that the motivation for positing the need for collective consent cuts against this account.

To overcome the challenges introduced, while maintaining the possibility of ethical xenotransplantation, some scholars have proposed a third response: the collective consent that is required is hypothetical (Rothblatt, 2004, 2023; Sparrow, 2009). On this view, risky-to-general-population research is ethically defensible if such research would be consented to by members of the general public if they were asked. Stated differently, the risks posed to the general public by xenotransplant research are justified because the general public would consent to it if given the opportunity. This solution requires no actual consent to be obtained, so the challenges of how to go about obtaining consent and who should do it are avoided. But hypothetical consent is not actual consent or consent based on a surrogate decision-maker's "substituted judgment" (what the subject would choose according to her accurately represented values), and it is those forms of consent that matter in research ethics. Researchers cannot place individuals at risk on the grounds that they would have consented had they been given the opportunity to do so without consulting anyone who is in a position to know better, and it is difficult to imagine why someone would agree to be placed at such risk with no direct benefit. If research places you at risk with no known offsetting benefit, why would you consent to it? To secure the hypothetical consent of those who do not stand to directly benefit from xenotransplant research, then, incentives or structural changes would have to be promised. For instance, access to healthcare for everyone and improved medical resources for vulnerable individuals and populations, those who stand at most risk by swine-derived zoonosis, might ensure hypothetical consent. But, as Sparrow (2009) argues, obtaining hypothetical consent in this way would exploit the already vulnerable: in exchange for their hypothetical consent, researchers would be offering them something they need and should have but lack. Implicit in this hypothetical offer, of course, is the threat that if they do not hypothetically consent, then they will not get something they need and should have.

At this juncture, it is important to examine the underlying principle motivating claims for collective consent in xenotransplantation: collective nonmaleficence overrides beneficence to an individual in the absence of collective consent (Bobier et al., 2025). The negative duty to do no harm to the many outweighs the positive duty to help the few. To formalize this thought, Rothblatt offers the following principle: "The consent of all parties that may be adversely affected should be obtained (Risk Principle)" (2004, p. 141). Yet, strictly speaking, this principle is not unconditionally true. Researchers do not obtain the consent of animals, since they cannot consent; yet, lots of animal research adversely affects animal subjects. Unconscious patients in the Emergency Department may undergo risky procedures without having offered consent. Finally, many actions performed now may or will adversely affect future non-presently-existing persons, but their consent cannot be obtained since future persons do not presently exist. To avoid these concerns, let us stipulate that this principle applies to existing persons who have decision-making capabilities, have reached the age of legal majority for medical decision-making, and whose intentional actions pose risk to other persons.

This principle requires that one person should not place other persons at risk without first obtaining their consent. This, in turn, entails that no one has the right to risk the life or well-being of others without their informed consent. This may be too strong, however. A little reflection reveals that we are placed at risk on a regular basis without having given consent. Living near a dam, a bridge, train tracks, a highway, or a nuclear power plant can pose a potentially significant risk, and yet, the affected people, apart from the democratic process in which they may or may not participate, are not owed an opportunity to individually consent or not to the dangers posed which would settle the question about what to do. Moreover, there is good reason to deny the claim that collective consent should be pursued whenever there is third-party risk. Challenge trials involve healthy volunteers receiving an experimental vaccine and then being exposed to a communicable disease—say, a virus—to test the efficacy of the vaccine. Exposure of a person to the virus may place their contacts post-exposure at risk without their consent, and thus, while third-party consent is not obtained, challenge studies are considered by many to be ethical. Of course, challenge trials are contentious, so to make our point more salient, consider the following case:

> Imagine that doctors have a miracle pill that can save a person from immediate death. This pill is currently expensive and is used only in wealthy cities in wealthy countries, so it is not currently widely available. The hope is that in time it will be cheap and available in generic versions the world over. There is another downside—everytime this miracle pill is administered there is a non-zero chance that 10 people in the world will get a mild cold. No one knows how or why this happens, only that it does.
>
> (Bobier et al., 2025)

Should doctors administer this miracle pill? Mild colds are largely benign for most people, albeit frustrating; however, they can lead to serious medical complications in medically vulnerable individuals. In a worst-case scenario, a cold can lead to death. While ten people suffering a mild cold is ethically salient, it seems that the risk-benefit favors administering the pill to an individual facing immediate death in the absence of the informed consent of all: even though a provider knows that ten random people will get a mild cold, he or she should still save the individual's life by administering the pill. The potential collective harm does not outweigh individual benefit in the absence of collective consent.

It might be objected that challenge studies and the miracle pill case present us with an acceptable, "minimal" risk level, however, that is defined (e.g., not greater than the risks faced in the ordinary course of life). If the miracle pill resulted in a nonzero chance that ten people in the world will get yellow fever, which poses a significant health risk, then it would seem as though providers should not administer the pill to individuals in the face of immediate death. Accordingly, collective consent should be pursued whenever there is potentially significant third-party risk. But there is good reason to deny this claim as well. A military weapons researcher who is designing a next-generation tactical missile to be used by the army should not be required to obtain the consent of those against whom the missiles might be used. A virologist researching a novel Ebola strain in a level 4 biosafety laboratory need not seek the consent of neighboring populations. These examples are particularly striking because they present potentially significant risk to other people but collective consent, the consent of all who may be harmed as a result of the research, need not be necessary for the research to be ethical. We do not mean to suggest that these activities are free from ethical scrutiny because consent from third parties is not needed. Surely, they can and should be regulated according to principles and considerations involving nonmaleficence and justice. Our point is only that both the weapons researcher and the virologist may be ethically justified in their research, assuming they pass other moral tests, in the absence of collective consent.

Long-term biosurveillance

There is a third response to the public health risks associated with xenozoonoses: xenograft recipients must undergo long-term biosurveillance. That is, in order to safeguard public health, there must always be an aspect of the clinical trial from which xenograft recipients cannot withdraw, either for a long time or in perpetuity. The argument for this position is that researchers have an obligation to minimize third-party risk when conducting research: the risks to society should be, at the very least, proportional to the potential benefits. If there is research that promises immense potential benefits to society, but also potentially significant public health risk (e.g., zoonosis), then a public health intervention that interferes with research participant liberty may be justified. To allow xenotransplant research participants to withdraw from follow-up monitoring, and thereby increasing the possibility of a pandemic, is irresponsible; the risk to the public is great and researchers ought to minimize third-party risk. This means safety measures that bring potential zoonotic risk to an acceptable level should be implemented. The options of doing so include long-term monitoring, long-term mandatory monitoring, or quarantine. To require quarantine would be excessively burdensome and overly restrictive of participant freedom, especially since participants pose only a possible risk to others. On the other hand, nonmandatory long-term monitoring would allow recipients to withdraw from monitoring, and this would hinder the prompt detection and management of a potential xenozoonosis. Requiring participants to waive their right to withdraw from follow-up, long-term monitoring is the least burdensome measure to protect public health and thereby ensure a favorable risk-benefit outcome (Bach et al., 1998; Reese et al., 2023; Iltis et al., 2024).

This is a compelling argument, but it is not without criticism (Bobier et al., 2024). To begin with, this proposal conflicts with an ethical tenet of human subject research, namely, that subjects have the right to withdraw without penalty (Code of Federal Regulations, 2023). This right has been affirmed by organizations from across the world, and it is easy to understand why:

1. Research may involve the intimate use of one's body and individuals have the right to bodily autonomy and integrity, viz., they get to choose what happens to their body (Capell et al., 2021; Lynch, 2020). If a person does not want something to happen to her body, others should respect that person's wishes. Rare exceptions aside, researchers cannot force people to do something they do not want to do.
2. The right to withdraw protects patients from undue influence and information asymmetry (Holm, 2011; McConnell, 2010). If a patient agrees to undergo a medical procedure about which they do not fully understand the impact of, they should be allowed to withdraw upon learning the impact more fully. Or, if it comes to light that a patient was coerced into a study, the patient should be given the opportunity to withdraw.
3. The right to withdraw protects patients from undue harm (Lynch, 2020; McConnell, 2010). For instance, forcing a patient to continue in an unwanted long-term study or procedure may be financially challenging for the patient or it can amount to a kind of imprisonment.
4. The right to withdraw protects patient privacy (Capell et al., 2021). If a patient does not want to disclose personal, perhaps compromising, information to a researcher, that person should be free to do so.
5. The right to withdraw has a role in promoting public trust in science (Schaefer and Wertheimer, 2011). Public trust in science is put at risk when people perceive scientists or law enforcement officials forcing healthy-looking people to undergo invasive, intrusive, and unwanted procedures.

Rights are beneficial for possessors, for they protect the possessor's interests from consequentialist considerations by generating *prima facie* obligations on others to respect them, and the right to withdraw is clearly beneficial for persons.

Some rights can be waived or transferred through a voluntary and informed decision. One waives their right to control what happens to a piece of property when it is sold to another through a valid transaction. One waives their right to control all aspects of a piece of property for a time one leases to another through a valid, time-limited contract. Others waive what they are otherwise entitled to in order to secure some other good they are entitled to or are free to pursue (e.g., documentation of informed consent for the sake of medical privacy). The question before us is this: can one "waive" one's right to bodily integrity and medical privacy in order to secure a potential benefit offered only in the context of clinical research? If we consent by waiving our rights, then is our right to control our bodies given up, suspended, or set aside for the sake of receiving the benefit?

Some appear to think so, that individuals can waive their right to withdraw. Applied to xenograft recipients, Entwistle and colleagues explain that "because of the public health implications of zoonotic disease, recipients may be required to waive their right to withdraw from infectious disease surveillance" (2022, p. 989). The argument is that requiring would-be xenograft recipients to voluntarily waive their right to withdraw respects their autonomy while protecting public health at the same time. More specifically, if would-be xenograft recipients are adequately informed about what lifelong biosurveillance requires and its importance to them and others, then they could voluntarily "waive" their right to withdraw in the sense that it is permanently suspended.

Two responses can be made, the first being that there is something curious about the argument in favor of waiving the right to withdraw from considerations of public health risk. As McConnell (2010) notes, if there is such a danger to the public, then individuals do not need to waive their right to withdraw, for in such cases, rights can be overridden. In general, when an individual poses a significant risk to others, we do not require them to voluntarily waive their rights; the state or other actors are justified in overriding their rights. One's right to life can be overridden by a police officer if they are actively threatening the life of another. So, if xenotransplantation poses such great risk, the risk is enough to override a patient's right to withdraw regardless of whether the patient has voluntarily done so. After all, if someone initially waives the right to withdraw and then reneges, the third-party risk justifies the invasive violation of autonomy.

Second, "waiving" the right to bodily integrity begs the question as to whether one's authority over what happens to one's body can be voluntarily suspended or given up. Indeed, the language of "waiving one's right to bodily integrity" involves a paradoxical view of autonomy insofar as it appeals to the autonomous choice of research participants to vacate their authority over what happens to one's own body and defer it to investigators until the investigators think it is appropriate to give back. Presumably, that autonomous choice is to be respected. Yet respecting this particular choice is to undermine the respectability of autonomous choice altogether since subsequent autonomous choices of the participant after the deferral need not be respected by the investigators if those choices are inconsistent with the investigator's goals. In other words, asking one to autonomously waive their right to bodily integrity is to treat what is to be respected as something that can undermine itself as an object of respect. In fact, the concept of consent as waiving one's rights is misleading, because it does not account for the dynamics of permission-giving that is grounded in *authority-retaining* consent. It is widely recognized in research ethics that informed consent is a process, not an event, that involves a give-and-take relationship with the participant with whom the researcher interacts at the participant's discretion. This is simply a function of the authority-retaining character of consent, which not only grounds respect for autonomous choice within biomedical ethics but also within sexual ethics and other domains, legal or otherwise, that involve rules against assault (Liberto, 2022).

Let us grant, for now, that an individual can waive her right to withdraw from biosurveillance. What happens if, at some future time, this individual reneges on the decision to waive her right to withdraw and decides to stop complying with biosurveillance protocols, whatever they may be? Let's say that after five years, a participant in a clinical trial for xenotransplantation is no longer willing to cooperate and complete any health surveys or undergo any invasive or noninvasive test that she originally consented to. Who exactly is going to "force" an uncooperative participant to have a blood test against their will? After all, this may require sedating or anesthetizing a competent individual and seems *prima facie* morally absurd. Arguably, *if* a clinical trial entails that mandatory long-term biosurveillance is necessary, then one must question whether the risk posed from beginning clinical trials can be morally justified to begin with.

Aside from the ethical challenges, there are pressing practical and legal challenges with long-term, lifelong biosurveillance enforcement. The best route would be to legislate an exception to the right to withdraw for xenotransplant research (Padilla et al., 2022). But practical questions abound. Who could issue such an exception and on what legal grounds? Who would enforce mandatory biosurveillance? On what grounds might officials legally enforce it? These are not insignificant questions, even though defenders of requiring recipients to "waive" their right to withdraw—already a dubious notion—tend to gloss over them. One suggestion would be to fashion consent to xenotransplantation along the lines of an employment-like contract (Lynch, 2014): in exchange for an organ, the recipient agrees to waive their right to withdraw. However, as Lynch (2020) notes, employment contracts assume bodily autonomy of both parties, since otherwise there would be forced labor: if employees are prevented from quitting, they are forced to work, which is illegal; accordingly, a US court is unlikely to enforce an employment-like contract in this context.

Another suggestion is a Ulysses Contract, which is sometimes used in psychiatry: patients with a psychiatric illness but who possess decision-making capacity and in remission may authorize their psychiatric team to provide a certain treatment even if they refuse at a later time (Reese et al., 2023; Spillman and Sade, 2007). Patients would be required to waive their right to withdraw from surveillance, with the form that they sign outlining, in the words of Spillman and Sade, "the surveillance schedule, as well as the proposed penalties for non-compliance" (2007, p. 270). But Ulysses Contracts (see Chapter 5 for a fuller analysis) in psychiatry are disanalogous to xenotransplantation for at least three reasons: (i) the goal of the advance directive in a psychiatric context is not applicable to a xenotransplantation context, (ii) enforcing Ulysses contracts in xenotransplantation is infeasible, and (iii) the ethical and regulatory hurdles such enforcement would require make binding enforcement impractical (Hurst et al., 2024; McConnell, 2010). Psychiatry patients enter into them for benefit to themselves, whereas xenograft recipients would enter into them for third-party benefit. In psychiatry, such a contract is freely entered into, whereas this would not clearly be the case in xenotransplantation: they are given the option of either agree or no organ. The contract is enforced when the psychiatric patient is manifesting a mental illness, whereas the contract may be enforced on non-ill xenograft recipients. Most importantly, a justification for Ulysses Contracts in psychiatry is that the patient's decision-making capacity is compromised, which is not the case in xenotransplantation. There is no legal precedent for forcing a healthy person with competent decision-making capacity to do something they otherwise do not want to do.

Conclusion

Public health concerns in the context of xenotransplantation have been raised for decades. They all stem from the ability for a microorganism to be transferred with the xenograft to the human recipient and be pathogenic in the human, affecting not only the recipient but possibly close contacts and the general public. Regardless of where one falls on the three proposals discussed in this chapter, we can all agree that continued research on zoonotic safety is advisable and that ethical xenotransplant research requires due diligence of both researchers and oversight organizations: the risk of zoonosis demands stringent caution, in design and in practice. What this looks like, however, is a matter for further discussion.

Let us take stock. We discussed current challenges in organ transplantation, specifically that there is a dearth of available organs for transplantation relative to the need (Chapter 1). We then examined the ethics of genetically engineering pigs to address the dearth of organs (Chapter 2) and the public health ethics posed by doing so (this chapter). Next, we will examine the ethical issues surrounding the selection of human recipients of genetically engineered pig organs (Chapter 4).

References

Bach, F. H., Fishman, J. A., Daniels, N., Proimos, J., Anderson, B., Carpenter, C. B., Forrow, L., Robson, S. C. and Fineberg, H. V. 1998. Uncertainty in xenotransplantation: Individual benefit versus collective risk. *Nature Medicine*, 4(2), 141–144.

Bobier, C., Hurst, D. J., Rodger, D. and Omelianchuk, A. 2024. Xenograft recipients and the right to withdraw from a clinical trial. *Bioethics*, 38(4), 308–315.

Bobier, C., Omelianchuk, A., Rodger, D. and Hurst, D. J. 2025. Collective consent to xenotransplantation: A critical appraisal. *Public Health Ethics*, 18(1), phae011.

Bourdain, A. 2001. *Typhoid Mary: An Urban Historical*. Bloomsbury Publishing USA.

Capell, W. H., Wynia, M. K., Hurley, E. A. and Bonaca, M. P. 2021. Should participants in clinical trials be able to withdraw from passive follow-up? *Ethics & Human Research*, 43(1), 32–36.

Code of Federal Regulations. 2023. *Common Rule*. Code of Federal Regulations.

Denner, J. 2016. How active are porcine endogenous retroviruses (PERVs)? *Viruses*, 8(8), 215.

Denner, J. 2021. Porcine endogenous retroviruses and xenotransplantation. *Viruses*, 13(11), 2156.

Denner, J. 2022. Virus safety of xenotransplantation. *Viruses*, 14(9), 1926.

Egerer, S., Fiebig, U., Kessler, B., Zakhartchenko, V., Kurome, M., Reichart, B., Kupatt, C., Klymiuk, N., Wolf, E., Denner, J. and Bähr, A. 2018. Early weaning completely eliminates porcine cytomegalovirus from a newly established pig donor facility for xenotransplantation. *Xenotransplantation*, 25(4), e12449.

Entwistle, J. W., Sade, R. M. and Drake, D. H. 2022. Clinical xenotransplantation seems close: Ethical issues persist. *Artificial Organs*, 46(6), 987–994.

Fischer, K. and Schnieke, A. 2022. Xenotransplantation becoming reality. *Transgenic Research*, 31(3), 391–398.

Fovargue, S. 2011. *Xenotransplantation and Risk: Regulating a Developing Biotechnology*. Vol. 14. Cambridge University Press.

Fovargue, S. and Ost, S. 2010. When should precaution prevail? Interests in (public) health, the risk of harm and xenotransplantation. *Medical Law Review*, 18(3), 302–329.

Holm, S. 2011. Withdrawing from research: A rethink in the context of research biobanks. *Health Care Analysis*, 19, 269–281.

Hurst, D. J., Padilla, L., Schiff, T. and Parent, B. 2024. Revisiting the use of Ulysses contracts in xenotransplantation. *Transplantation*, 108(2), 369–373.

Iltis, A. S., Silverman, H. J. and Sade, R. M. 2024. Research ethics: Must subjects waive the right to withdraw from a xenotransplant clinical trial? *The Journal of Thoracic and Cardiovascular Surgery*, 167(5), 1880–1884.

Johnson, L. S. M. 2022. Existing ethical tensions in xenotransplantation. *Cambridge Quarterly of Healthcare Ethics*, 31(3), 355–367.

Jorqui-Azofra, M. 2020. Regulation of clinical xenotransplantation: A reappraisal of the legal, ethical, and social aspects involved. *Xenotransplantation: Methods and Protocols*, 315–358.

Liberto, H. 2022. *Green Light Ethics: A Theory of Permissive Consent and Its Moral Metaphysics*. Oxford University Press.

Łopata, K., Wojdas, E., Nowak, R., Łopata, P. and Mazurek, U., 2018. Porcine endogenous retrovirus (PERV)–molecular structure and replication strategy in the context of retroviral infection risk of human cells. *Frontiers in Microbiology*, 9(730), 1–11.

Lynch, H. F. 2014. Human research subjects as human research workers. *Yale Journal of Health Policy, Law, & Ethics*, 14, 122.

Lynch, H. F. 2020. The right to withdraw from controlled human infection studies: Justifications and avoidance. *Bioethics*, 34(8), 833–848.

McConnell, T. 2010. The inalienable right to withdraw from research. *Journal of Law, Medicine & Ethics*, 38(4), 840–846.

Menache, A. 1999. Stop now before it's too late. *Bulletin of the World Health Organization*, 77(1), 76–78.

Nellore, A., Walker, J., Kahn, M. J. and Fishman, J. A. 2022. Moving xenotransplantation from bench to bedside: Managing infectious risk. *Transplant Infectious Disease*, 24(6), e13909.

Niu, D., Wei, H. J., Lin, L., George, H., Wang, T., Lee, I. H., Zhao, H. Y., Wang, Y., Kan, Y., Shrock, E. and Lesha, E. 2017. Inactivation of porcine endogenous retrovirus in pigs using CRISPR-Cas9. *Science*, 357(6357), 1303–1307.

Padilla, L. A., Hurst, D., Maxwell, K., Gawlowicz, K., Paris, W., Cleveland, D. and Cooper, D. K. 2022. Informed consent for potential recipients of pig kidney xenotransplantation in the United States. *Transplantation*, 106(9), 1754–1762.

Ravelingien, A. 2007. Xenotransplantation and the harm principle: Factoring out foreseen risk. *Journal of Evolution & Technology*, 16(1), 127–142.

Reese, P. P., Gelb, B. E. and Parent, B. 2023. Unique problems for the design of the first trials of transplanting porcine kidneys into humans. *Kidney International*, 103(2), 239–242.

Rodger, D., Hurst, D. J. and Cooper, D. K. 2023. Xenotransplantation: A historical-ethical account of viewpoints. *Xenotransplantation*, 30(2), e12797.

Rothblatt, M. 2004. *Your Life or Mine: How Geoethics Can Resolve the Conflict Between Public and Private Interests in Xenotransplantation*. Routledge.

Rothblatt, M. 2023. Geoethics and xenotransplantation. In *Xenotransplantation: Ethical, Regulatory, and Social Aspects*. Cham: Springer International Publishing, 69–74.

Sade, R. M. and Mukherjee, R. 2022. Ethical issues in xenotransplantation: The first pig-to-human heart transplant. *The Annals of Thoracic Surgery*, 113(3), 712–714.

Salyer, S. J., Silver, R., Simone, K. and Behravesh, C. B., 2017. Prioritizing zoonoses for global health capacity building—themes from one health zoonotic disease workshops in 7 countries, 2014–2016. *Emerging Infectious Diseases*, 23(Suppl 1), S55.

Schaefer, G. O. and Wertheimer, A. 2011. Reevaluating the right to withdraw from research without penalty. *The American Journal of Bioethics*, 11(4), 14–16.

Sparrow, R. 2009. Xenotransplantation, consent and international justice. *Developing World Bioethics*, 9(3), 119–127.
Spillman, M. A. and Sade, R. M. 2007. Clinical trials of xenotransplantation: Waiver of the right to withdraw from a clinical trial should be required. *Journal of Law, Medicine & Ethics*, 35(2), 265–272.
Veatch, R. M. and Ross, L. F. 2015. *Transplantation Ethics*. Georgetown University Press.
Xie, Z., Pang, D., Yuan, H., Jiao, H., Lu, C., Wang, K., Yang, Q., Li, M., Chen, X., Yu, T. and Chen, X. 2018. Genetically modified pigs are protected from classical swine fever virus. *PLoS Pathogens*, 14(12), e1007193.
Yang, L., Güell, M., Niu, D., George, H., Lesha, E., Grishin, D., Aach, J., Shrock, E., Xu, W., Poci, J. and Cortazio, R. 2015. Genome-wide inactivation of porcine endogenous retroviruses (PERVs). *Science*, 350(6264), 1101–1104.

4 Who should receive a xenotransplant?

To bring a therapeutic medical intervention to market, the intervention must go through various stages of experimentation, and at each stage of the long process, researchers have to recruit participants. Given the well-known abuses of research participants throughout history, ethical guidelines exist to protect them. The ethical questions to be settled are quite complex. For instance, does research on the terminally ill incur greater risks of exploitation than on those who are not terminally ill? What role could healthy volunteers play, if any? How could the process of informed consent mitigate concerns of exploitation? At what point does a participant's desperation become an undue influence on decision-making? What inclusion criteria are defensible? Should children be considered? If so, when? Would the study's design yield valuable scientific information? Does the risk to the participant outweigh the social benefits of the research? The ethical issues do not end at the clinical stage, for bringing a therapeutic intervention to market raises a host of other ethical issues surrounding distribution. For instance, how should the intervention be distributed? Will the costs exclude and further marginalize the underprivileged? What role should the government play in the intervention's distribution? If a new intervention is less optimal than an existing intervention but more cost-effective, which should the clinician recommend?

Let us assume that xenotransplantation is deemed ethically justified, and that the animal and public health considerations discussed in the preceding chapters are favorable to proceeding. In this chapter, we engage with ethical questions around the issue of who should receive a xenotransplant. Currently, as xenotransplant research is in early clinical trials, this is a question about research participant selection: who should be enrolled in xenotransplant clinical trials? Assuming xenotransplant research progresses to clinical practice, the question becomes, who should receive one? If so, when? As will be clear from above and in what follows, these are distinct questions and will be treated as such. We proceed in order, beginning

DOI: 10.4324/9781003472223-5

with xenotransplant research participant selection and then xenotransplantation as therapeutic intervention.

Before beginning, we should note that transplant medicine is marked by stories of desperate patients asking their innovative doctors to try anything. Most of them are tragic, ending in failure, but also with knowledge gained that sets others up for success later. These successes involved patients who were similarly in need of an organ and were willing to take a chance with an untested intervention. The morality of these events is complex and worthy of commentary, but the more acceptable ones reflect an understanding between the doctors and the patients that things might not work out but that knowledge to help others might be gained. The less acceptable ones appear to lack this understanding among other things. The ethical issues involved go beyond the elements of informed decision-making, however, so we organize our approach somewhat artificially under the headings of "research" and "therapy" knowing these categories are not always neatly separable in the doctor–patient relationship in experimental transplant medicine.

The ethics of human research

Obviously, human subject research should be ethical, but what makes for ethical research, however, is less obvious. To help focus the discussion, we appeal to Ezekiel Emanuel and colleagues' (2000) influential article on clinical research in which they set out seven conditions for ethical human subject research:

1. *Social value*: The clinical trial must propose enough potential social benefit to justify the potential risk to participants: "Only if society will gain knowledge, which requires sharing results, whether positive or negative, can exposing human subjects to risk in clinical research be justified" (2000, p. 2703).
2. *Scientific validity*: The clinical trial should be scientifically rigorous in design, which minimally involves valid methods and practical feasibility.
3. *Fair subject selection*: The clinical trial should solicit participants solely on the basis of the scientific goals of the study and the burden of participation should be, as far as possible, shared among all who may benefit.
4. *Favorable risk-benefit ratio*: The potential benefits to participants *and* society should outweigh the risks to the participants and society. Researchers must minimize risk to participants and, as they are able, maximize potential benefit. If participants have little potential benefit, then the societal benefits should outweigh the potential risks.

5. *Independent review*: Research should be subject to independent review by the institution's IRB and other oversight committees.
6. *Informed consent:* Participants "must be accurately informed of the purpose, methods, risks, benefits, and alternatives to the research; understand this information and its bearing on their own clinical situation; and make a voluntary and uncoerced decision whether to participate" (2000, p. 2706).
7. *Respect for potential and enrolled subjects*: Participants should be treated with respect, which minimally includes the maintenance of privacy and confidentiality.

We take these conditions to identify a baseline for ethical research, such that if any one of them is *clearly* compromised, then so too is the overall research proposal. The "clarity" with which a condition has been breached is a function of the condition's dependence on values that come in degrees. A proposal can be more or less socially valuable, the study can be more or less rigorous, the subject selection more or less fair, the risk-benefit ratio more or less favorable, the consent more or less informed, and the subjects more or less respected. Even the review can be more or less independent given the politics that influence institutional review boards. We note these shifting sands because they are matters of interpretation inherent to any evaluation and do not always admit of a clear answer.

The issue we focus on is whether xenotransplant clinical research can adequately satisfy the fair subject selection and favorable risk-benefit ratio criteria. To simplify the argument of this chapter, we will assume for the sake of argument that the value of xenotransplant clinical research is high, that the research is rigorously designed, that informed consent is obtainable, that recipients are treated with respect, and that the research undergoes independent review. Many of these assumptions are discussed elsewhere in this book.

Limiting trials to those with no alternatives: is it fair?

Xenotransplant clinical trials are in their beginning stages, and have much to consider as researchers and organizations over the span of decades have issued guidance on potential participant inclusion and exclusion criteria. In general, they stipulate five conditions for enrolling participants in xenotransplant clinical trials. The patient is (a) seriously ill (the Serious Illness condition); (b) lacks safe and effective therapeutic alternatives (the No Alternatives condition); (c) stands to benefit from trial participation (the Individual Benefit condition); (d) understands risks and consequences and voluntarily consents (the Understanding condition); and (e) is motivated to comply with follow-up (the Adherence condition). Some of these conditions will be explored in other chapters such as the Understanding condition, which enables participants to make a free and informed decision to participate (Chapter 5), and the Adherence condition, which captures the importance of monitoring for xenozoonosis and continued treatment (Chapter 3).

The Serious Illness, No Alternatives, and Individual Benefit conditions reflect the observation that initial xenotransplant clinical trials will likely involve patients who are unlikely to receive an allotransplant and who are ill. These conditions are found in multiple statements from research authorities. The United States Food and Drug Administration (FDA), for example, posits the following:

> Due to the potentially serious public health risks of possible zoonotic infections, you should limit xenotransplantation to patients with serious or life-threatening diseases for whom adequately safe and effective alternative therapies are not available, except when very high assurance of safety can be demonstrated. You should limit candidates to those patients who have potential for a clinically significant improvement with increased quality of life, following the procedure. You should also consider the patient's ability to comply with public health measures as stated in the protocol, including long-term monitoring.
>
> (2016, X.D.43)

Similar guidance is offered by the World Health Organization's (WHO's) *Changsha Communiqué*:

> Investigators should select trial participants for whom there is no adequately effective alternative therapy available [highlighted by the authors] and who understand the risks and consequences of the procedure, including the need for compliance with life-long follow up and who are motivated to modify their behaviour accordingly.
>
> (2008, p. 3.3)

While the information gained from these trials will potentially benefit society in the future, participants can expect little benefit themselves. For this reason, the Nuffield Council on Bioethics' *Animal-to-Human Transplants: The Ethics of Xenotransplantation* observed that it "would be hard to justify posing any potential public health risk unless the first xenografts were used to save the lives of people with no alternative possibilities of treatment" (1996, p. 7.11). More recently, the American Medical Association's *Code of Medical Ethics* advises that "recruitment is restricted to patients with serious or life-threatening conditions for whom no adequately safe and effective alternative therapies are available unless there is documented, very high assurance of safety" (2017, p. 6.3.1D).

Terms like an "adequately safe and effective alternative" and "documented, very high assurance of safety" (phrases echoed throughout guidelines) are unspecified, but a common interpretation of these phrases is that there is sufficient laboratory and animal experimentation data that justifies an experiment in humans, and there is no "tried and true" alternative to xenotransplantation, for example, allotransplantation or supportive therapies such as dialysis or LVADs. Reflecting these assumptions, the authors of a position paper on xenotransplantation research published in *The Journal of Heart and Lung Transplantation* in 2000 wrote that "patients with end-stage cardiac or pulmonary disease may represent ideal candidates *when no other therapy will support life*" (Cooper et al., 2000, p. 8.1). When discussing cardiac xenotransplantation, in particular, they suggest that a case can be made for including "patients who are in end-stage failure *but who are unacceptable for allotransplantation*" (8.2.1; emphasis original). The kind of patient needed, then, appears to be one who is too sick for standard therapy, but not so sick as to confound the study results or fail to receive some benefit from the xenograft.

As much as the No Alternatives condition enjoys widespread support, it invites scrutiny because "a lack of psychosocial support" or "a history of noncompliance" often disqualifies one from allotransplant candidacy. How these disqualifying features come to be attached to any given patient may be a result of bias (Berry et al., 2019). Regardless, if these sorts of patients are selected for xenograft research, they would be those that either lack social support or risk confounding the study because of a lack of compliance. Indeed, this was precisely the sort of patient who received the first porcine heart in 2022 (Griffith et al., 2022). Therefore, the selection criteria compromise fairness insofar as the burdens of the research fall on the "unsupported" or those who were subject to bias. Moreover, the rigor of the study may be harmed if the participants, who are deemed to have a history of nonadherence, do not adhere to the research protocol. Lastly, it is worth noting that the potential benefit to such participants is largely undefined probably because there is little to expect when one does not have the requisite support or the willingness to adhere.

Hence, some researchers have argued for expanding inclusion criteria to include patients who are still eligible for an allotransplant, but are so far down the waitlist that they will likely die or be delisted before receiving a human organ. These patients may be healthier than those who are not candidates for allotransplantation, but for one reason or another, will likely never receive an allotransplant. Openness to this possibility can be found in a paper by Richard Pierson III and colleagues who wrote that "Clinical trials should enroll 'patients lacking timely access to an allograft or who are expected to be served poorly by currently available therapeutic alternatives'" (2022, p. 1720). Likewise, the judgment of David Cooper and colleagues is that "these patients [for selection] should be significantly disadvantaged or completely excluded from being offered a human organ" (2017, p. 1552).

The rationale for broadening the inclusion criteria is rooted in the fact that even waitlisted individuals are at risk of dying and have little hope of a therapeutic alternative. They are just as much in need of a xenotransplant as those who, for whatever reason, are excluded from allotransplantation but who still need an organ. It is not just those who are excluded from allotransplantation that are at risk of death and who may benefit from participating in a xenotransplant clinical trial; those who are low on the transplant list also face the risk of death. Yet they have the requisite psychosocial support and are predicted to be adherent to the treatment regimen. Thus, the potential benefit of an experimental xenograft would be to "bridge" the patient to allotransplant, that is, buy more time for the patient languishing on the waitlist and increase their chance of receiving a human organ. Such individuals would, in the words of the FDA guidelines, more clearly "have potential for a clinically significant improvement with increased quality of life, following the procedure" (2016, X.d).

Expanding the criteria for those on the waitlist maximizes the risk ratio for satisfying Individual Benefit but perhaps places too much of a burden on the sick to share in the research burdens. What about expanding the criteria to healthy patients? Against the advice of the WHO, the FDA, the AMA, and the Nuffield Council is the suggestion that healthy subjects be solicited for xenotransplantation research. The proposal would be limited to clinical trials for renal replacement in which a person who has already decided to donate is approached about receiving a xenograft in place of their explanted kidney. Consequently, the volunteer would take on not only the risks associated with procurement surgery and living with only one kidney but also the known risks associated with taking antirejection medication and the unknown risks of sensitization and zoonotic infection with the associated duties to be monitored and studied.

Although this proposal is beset by study design challenges—most importantly, how to measure "xenograft function in participants with native kidney function" (Adams et al., 2024, p. 98)—the selection criteria do not discriminate on the basis of health status. The concern over therapeutic misconception (the idea that the purpose of the trial is curative not scientific) is mitigated since the proposal obviously provides no medical benefit to the participant. Only the positive experience of knowing one is contributing to scientific knowledge could motivate the choice. To be sure, there is a strong presumption against subjecting volunteers to non-beneficial invasive surgery, but it is not without precedent in the history of research ethics as placebo-style arthroscopic "sham" knee surgeries have been ethically performed to test the efficacy of standard therapies (Moseley et al., 2002).

Yet the proposed surgery is not a sham, because some of the risks of the laparoscopy are accepted for the sake of live donation. What needs to be determined is whether the charity of the donor might extend beyond helping a needy recipient to contributing to scientific progress.

Although offers of an implanted xenograft would likely be rejected, it may at least be worth asking for the sake of including a few individuals in an adequately designed study, for example, on how to best understand or treat the effects of donor-specific antibody-mediated rejection (Tector, 2024). The knowledge gained from these studies could even help mitigate the problem of therapeutic misconception among the seriously ill if that knowledge were to help increase the potential for therapeutic benefit. We suspect, however, that IRBs and investigators will be hesitant to subject healthy donors (and their close contacts) to the risks of zoonotic disease and the complications associated with immunosuppression (increased risk of cancer and opportunistic infection) to gain these scientific benefits. That is to say, pursuing a greater degree of fairness in the distribution of the research burden can decrease the benefit-risk ratio in some groups of participants on whom the research burdens are placed—"leveling down" some to achieve some sense of fairness with others can be hard to justify.

Fair subject selection

Given the complexity of determining whether the proposed inclusion-exclusion criteria are fair, we examine four principles—generalizability, shared burden, fair opportunity, and third-party risk—to assess the ethical defensibility of subject selection criterion under consideration.

Generalizability

The first principle is that participant selection should allow researchers to generalize to the broader community. This is a matter of fair inclusion. "Fair inclusion," McKay and Saylor write, "requires that the selection of participants be sufficiently inclusive to ensure that the research in question is appropriately generalizable to clinically distinct populations" (2020, p. 7). The rationale is straightforward: if clinical research enrolls only medically fragile individuals, then this may not produce knowledge that is applicable to other patient populations that would stand to benefit from a xenograft, namely, those who are not as seriously ill. Moreover, the inclusion of patients with different demographic identifiers associated with sex, race, and age increases the likelihood that the study results will contribute to the construction of a safe and effective clinical modality that will be broadly beneficial to the general patient population.

If xenotransplant participant selection continues to be limited to older adult males for whom an allotransplantation is not possible or highly unlikely and who are at risk of dying on account of being seriously ill, then it is questionable whether researchers can extrapolate meaningful clinical data to healthier individuals with different demographics.

Shared burden

The second principle, which we've appealed to a few times now, begins with the observation that it is unfair to ask already burdened individuals to become even more burdened when there are other people who are less burdened and can thereby more easily assume the burden. Practically, persons who are not so burdened are in a better position to bear the burden of participating in a clinical trial, as McKay and Saylor explain, "In the context of the cooperative project it is unfair to ask some individuals to bear excessive burdens for the purposes of benefiting others" (2020, p. 8). This echoes what is set forth in various guidelines. The *Belmont Report* classifies "the very sick" as a vulnerable group that is targeted for inclusion precisely because they have no other alternative, not because they are uniquely tied to the research (National Commission for the Protection of Human Subjects of Biomedical and Behavioral Research, 1979, p. 19). Medical research is a socially beneficial undertaking, yet it poses burdens to participants. Insofar as it is feasible, these burdens should be distributed equitably; if the burden of participation is high, then less burdened individuals should have priority in participation.

Institutional guidelines favor limiting trials to those who are most burdened of all—those who are seriously ill and at serious risk of death. The motivating assumption is that this population stands to benefit the most because they are thought to have "nothing to lose" (Cooper and Lanza, 2000, p. 179). But of course they do have something to lose if the study goes awry. "This [participation in the trial] may result," as Cooper and Lanza caution, "in the patient's suffering a drawn-out and painful death, rather than a relatively peaceful death with dignity" (2000, p. 178). They argue that those in whom the research "will do no harm"—and it is not clear how we could know this—ought to be included as well (2000, p. 179). If we could know that, then this would be more consistent with sharing the burdens of research across a wider patient population at least. In any event, the institutional guidelines appear to favor the limiting of participation in the research to the most medically vulnerable. If it is responded that xenotransplant research is too risky for candidates for allotransplantation, even candidates low on the allotransplant waitlist, then it would seem to be too risky for everyone. We risk burdening the already-burdened with something we deem too burdensome for the not-as-burdened. The burdens of bringing this research to the bedside would seem to be unduly placed on the shoulders of the most burdened.

Fair opportunity

The third principle is fair opportunity, which, as MacKay and Saylor explain, "requires that prospective participants be granted a fair opportunity to participate in research that is ex ante beneficial" (2020, p. 9). What this means, they write, is that "investigators must recognize that all who meet the scientific requirements of the study have a *prima facie* equal claim to participate" (2020, p. 9). This criterion is grounded in the recognition of moral equality among people, which entails the equal consideration of interests. For example, if otherwise stable people in need of an organ, regardless of their age, may benefit from xenotransplant research, and there is no reason inherent in the research itself why they should be excluded, xenotransplant guidelines that would limit research to adults only would violate the fair opportunity criterion. That is to say, pediatric patients would be unfairly excluded. Everyone who could benefit from a xenotransplant should be allowed to participate.

It must be observed that the fair opportunity principle is a *prima facie* equal claim, and there are compelling reasons to limit the opportunity to participate along the lines of five conditions above. The risk-benefit ratio favors the exclusion of those with a proven and available therapeutic alternative. But for those who lack a therapeutic alternative, and who are at risk of dying, the risk-benefit ratio is more favorable to participating. We discuss differential risk-benefit ratios later in this chapter.

Third-party risk

The fourth principle set out by MacKay and Saylor regards third-party risk and "demands that participants be selected to ensure a fair distribution of risks to bystanders" (2020, p. 5). The rationale for this condition is that the selection of participants "plays a role in the creation and distribution of these third-party risks" and there are reasonable limits to place on third parties (2020, p. 10).

The case can be made that this condition is unsatisfied for reasons already mentioned: xenotransplant research carries a low but theoretically plausible risk of zoonotic transmission from the xenograft recipient to close contacts and to the wider public. This public health concern around emerging zoonotic diseases has been heightened following the COVID-19 pandemic, and there is a renewed interest in how they can be prevented (Petrovan et al., 2021). It is therefore essential that patients and the public are sufficiently confident that everything possible is being done to keep this risk as low as possible. Alternatives to xenotransplant research such as policy changes and increased donor awareness do not pose such a serious third-party risk.

Risk-benefit concerns

As already mentioned, there is thought to be a compelling reason to limit initial xenotransplant clinical trial participation to individuals who are Seriously Ill and have No Alternatives: they have "nothing to lose" as it were, and can always transition to comfort measures when the going gets tough. Trial participation for them is thought to offer a more favorable risk-benefit ratio than for individuals who are less ill, not at risk of dying, or have a therapeutic alternative. Emanuel and colleagues explain that, in general, would-be participants "who are eligible based on the scientific objectives of a study, but are at substantially higher risk of being harmed or experiencing more severe harm, should be excluded from participation" (2000, p. 2704). Xenotransplant clinical trials will be invasive and participants will have to accept both known and unknown risks that include infection, organ rejection, and death. Research participation is unfavorable for those who have a proven and available therapeutic alternative, but for those who lack such an alternative, participation may prove beneficial. Or so the argument goes.

The most vulnerable patients in this class are children with congenital cardiac conditions. The "Baby Fae" experiment, which used baboon cardiac graft, was motivated by the hopelessness caused by the fatal condition of hypoplastic left-heart syndrome. As Arthur Caplan argued at the time when her case was being discussed, the grim fate of such similarly situated patients "would appear to justify allowing their participation in further clinical trials to advance scientific understanding of the feasibility of xenografting when appropriate ethical and scientific requirements have been met" (1985, p. 3342). Debate continues about when those requirements might be met. Ethically, it is preferable to develop the intervention in adults first unless it cannot be done so in them. Scientifically, at the time of this writing, there is a lack of preclinical data for a porcine-to-human child experiment.

One might also wonder whether, despite every attempt to minimize risk and promote participant benefit, the risk-benefit ratio is favorable to participants who are seriously ill and have no alternatives. The case can be made that participation in initial xenotransplant trials would be a *bad deal* for the participants: they will undergo invasive surgery and still likely die shortly thereafter. While their death due to organ failure is likely unavoidable, a medically burdensome death is not. This, again, raises the issue of whether hospice care would be better for them. Better to die as peacefully as possible rather than undergo an invasive, harmful operation.

The response to this is to argue that the risk-benefit ratio is more favorable, albeit slightly at best, for medically fragile persons than for others.

"The risk–benefit assessment," Agrawal and Emanuel write, "requires consideration of the available alternatives," and so, if there is no therapeutic alternative, then "a slight chance of therapeutic benefit is not unreasonable" (2003, p. 1077). In addition, there may be some benefits to participating in initial xenotransplant trials, including collateral psychological benefits, greater access to healthcare resources, prolonged life, or improved quality of life.

Even if initial xenotransplant participants are seriously ill and have no alternatives and a poor risk-benefit ratio, their poor ratio is not enough to justify forbidding research. The impetus for the research is that it promises significant potential social benefit. Although the initial participants will very likely die shortly after undergoing this invasive surgery, as the first four recent xenotransplant recipients did, the primary ground for permitting this research is the societal benefit of increased scientific knowledge. Hence, early xenograft research will likely be framed as benefiting society more than benefiting the individual participant.

The interest in maximizing individual and social benefits raises the concern of potential exploitation, however. Commenting on the consent process in the Baby Fae case, Knoll and Lundberg (1985) found that the expected individual benefits of the xenograft were overstated. The opportunity to exaggerate the benefits of xenotransplantation is greater when the patient has been deemed to have "nothing to lose." So, exploitation, which occurs when one person takes unfair advantage of another person, is a risk. A common way of conceptualizing exploitation is that A exploits B when A takes advantage of B to gain unfair benefit (Wertheimer, 2008; Snyder, 2012). Some scholars add one more condition to this account, namely, a vulnerability clause: A gains excessively from B precisely because of B's vulnerable or precarious circumstances (Liberto, 2014). For example, if B is drowning in the ocean and A is the only person around who can easily help B, it would be exploitative of A to offer to save B only if B pays $500. Overlooking whether A has a duty to save, A takes unfair advantage of or exploits B's precarious situation. Exploitation may be procedural (e.g., A benefits from B without B's consent or through deception) or may be a function of the outcome (e.g., either B is harmed or A gets more than he or she is entitled to get from B). Exaggerating the benefits of xenotransplantation in the consent process would be an instance of the procedural variety, and garnering fame and fortune at the expense of the moribund participant is a function of the outcome. At least, that is how it can be argued.

Do trial requirements that limit participation to individuals who are seriously ill with no alternatives risk exploitation? The initial participants are medically vulnerable and unlikely to personally benefit from participating in clinical trials for xenotransplantation. So, the ethical permissibility of initial clinical trials is primarily grounded in a favorable social benefit outweighing the personal risk. Researchers thus appear to burden these participants with an unfavorable-to-them trial for the sake of knowledge that may benefit others in the future. The *Belmont Report* classifies "the very sick" as a vulnerable group, and they are targeted for inclusion precisely because they have no other alternative, not because they are uniquely tied to the research—candidates for allotransplantation or even healthy living donors can be included but will not be. So, it can be argued that selection criteria based on the guidelines are exploitative: participants are likely to gain very little personal benefit and are selected because of their medical vulnerability, while researchers and society potentially gain a lot.

Yet it can be responded that participants should be free to enter into research, and that if someone volunteers to participate and gives their informed consent, they are not being taken unfair advantage of. If such individuals give informed consent, as there is little reason to suppose their decision-making capacities are compromised, then their participation in initial xenotransplant research can be justified. Because pursuing comfort measures is a reasonable option, it is not the case that the patient cannot reasonably say no to being in the trial. Hence, the offer to participate is neither an instance of coercion nor undue influence. Besides, the participants may benefit more than we expect. The success of this response assumes (a) that informed consent to participate is possible and (b) that informed consent eliminates concerns of exploitation. To exclude the opportunity to participate in such research seems paternalistic and unjustified. However, note that this response is also available to healthy live kidney donors who may desire to contribute to the research.

Clinical xenotransplantation

The goal of xenotransplant research is for it to become a clinical reality similar to the efficacy of allotransplantation in which 80–90% of the recipients survive for longer than one year. At least, that is the goal. Many scientific advances need to happen to achieve that goal, but if it is achieved, we have good reason to believe that the ethical issues will be related to resource problems that involve scalability, affordability, and equitable distribution. Xenografts will be a commodity sourced by industry suppliers who will seek a lucrative return on their investment. Private insurance companies and government healthcare funders will have to decide on what extent, if any, they will cover the costs. Industries will market their grafts, perhaps directly to "consumers," as being superior to others. Surgeons will have to navigate the complicated task of comparing outcomes from competing sources. Ultimately, transplant teams will bear the responsibility of recommending a treatment plan that will be best for the patient, which will involve comparing and contrasting xenotransplantation with allotransplantation and whatever other alternative therapies there may be.

To help focus this responsibility, let us suppose that allotransplantation will continue to be the "gold standard" because allografts pose no risk for zoonotic disease and enjoy a much more solid evidence base. Let us also suppose that the most likely clinical application in the near future will be bridging patients to allotransplant, something that could be especially useful for patients with liver disease. The clinical application would be to use the porcine liver to provide a "temporary detox" to a patient whose native liver needs time to heal or for a human liver to become available (Mallapaty, 2024). The clinical goal of this intervention is to reduce morbidity and mortality associated with liver failure in the acute phase of treatment; the duration of use is relatively short (e.g., up to ten days), and the risks of hyperacute rejection and zoonotic infection are clinically managed in the critical care setting. The goal is specific, the time of use is definite, and the risk factors are closely monitored. This sort of clinical application, especially if a hepatectomy is unnecessary, could become standardized while ongoing research with other xenografts is conducted for more complex surgical interventions with broader goals of care that require longer usage over an indefinite amount of time that is subject to less stringent outpatient monitoring.

This type of incremental experimental-to-clinical pathway raises the question of relating therapy and research more closely. While traditional bioethics, following the Belmont Report, has traditionally separated therapy from research in order to avoid conflicts of interest and the abuse of participants, the history of transplant medicine has always involved a closer relationship that tries unproven interventions on patients seeking therapy. It is hard to imagine how the field could have progressed without this more integrated approach. Fox and Swazey's (1974) sociological portrait of the early years of transplant medicine was that of a surgeon-researcher who not only had, in the their words, "the courage to fail" but also the willingness to enter into long-term relationships with their patients, both donors and recipients, in which ethical choices had to be made about whether to continue with standard therapy (e.g., dialysis), pursue transplant, intervene in the context of a rare condition, or satisfy research needs. Therefore, it may be useful to explore models that integrate research with therapy so as to maximize the benefits for patient outcomes and generalizable knowledge for other researchers. A registry of xenotransplants that have clearly defined therapeutic goals would help other researchers capture relevant epidemiological data about the recipients, the organ source, and the clinical interventions deployed to help expand relevant therapeutic possibilities and training for other teams.

What progress might look like under this model is described in detail by Cooper and Lanza (2000). Starting with the "temporary detox" example of the extracorporeal porcine liver, testing the length of time the liver could be used or whether multiple livers could be used would occur in other cases. Knowledge gained from these interventions could also be shared with researchers developing "artificial liver" systems to be used for extracorporeal hemoperfusion. The goal would be to determine if the extracorporeal organ would be worth implanting in patients with acute liver failure whose native livers cannot be salvaged. Once enough confidence is garnered, the first porcine liver transplant could occur to relieve the patient of multiple detox sessions and extend time on the "bridge" to allotransplant.

A feature (not a bug) of this pathway is that xenografts are not the only clinical modality available to the patient. Whatever xenogeneic extracorporeal liver perfusion might promise, the outcomes of using it have to be measured against those produced by standard intensive care management. Shortly after Cooper and Laza's book was published in the year 2000, a meta-analysis found that the extracorporeal technique of using human or pig livers "did not surpass the success of conventional intensive care treatment" (Pascher et al., 2002, p. 309). Be that as it may, the point is that there could be a clinical application for xenografts that mixes research and therapy from which transplant teams can both learn and pursue the best treatment options available. In any event, the clinical goal of xenotransplantation is to be a bridge to allotransplantation while knowledge is gathered for establishing it as an alternative therapy on par with allotransplantation.

This approach provides a framework for making listing decisions. Receiving a xenograft would be part of what qualifies one to be waitlisted in the "bridge therapy" phase (the acceptable scenario); delisting would occur only if the patient critically decompensates (worst-case scenario) or is meeting benchmarks comparable to allotransplantation (best-case scenario). One benchmark that will no doubt cause consternation and uncertainty will be the risk of contracting zoonotic infection. Because allotransplantation does not have this problem, it may be the case that xenotransplantation will only ever be a "bridge" and never a "destination." However, we can only speculate. Perhaps one's priority on the list will be lowered if one is doing well; perhaps it will be raised if not. Perhaps the time predicted for the "destination" therapy will be relatively short and closely monitored. Only time and experience will tell what is truly possible.

Arguments for "destination" therapies will most likely come from those advocating for patients in renal and cardiac failure who are not candidates for allotransplantation, but are able to receive alternative therapies. Although it will be difficult to justify working up a patient for a cardiac xenograft when an LVAD is available, patients might understandably reject the LVAD based on certain "quality of life" considerations. Depending on a battery-powered device with a drive-wire protruding from the chest or abdomen that must remain free from infection and moisture may be unacceptable to a patient who would rather take a chance with the pig organ. The same sort of patient-centered "quality of life" considerations might result in a rejection of ongoing dialysis sessions for renal candidates. The clinical conversation will then have to be about the comparative "quality of life" on comfort measures versus one with an experimental xenograft. One type of patient Cooper and Lanza describe, that is, one who is elderly with "severe and widespread atherosclerotic disease," and "with advanced diabetes" and who was "turned down for a place on the transplant list," is just the sort who would most benefit from a sober disclosure about the risks and benefits of each option (2000, p. 183). Yet it may be questionable to frame the xenograft as a "destination" *therapy* if the benefits are meager and the scientific knowledge to be gained is substantial. Offering it under such a description is beset by conflicts of interest and the consent process would be corrupted by therapeutic misconception. Correcting for these problems would be to more clearly classify the option as experimental with the intention to meet the ethical standards for clinical research.

Conclusion

Who shall go first? This question is important, although the answer is debatable on scientific and ethical grounds. As we inch closer to initial trials for other types of xenotransplantation beyond kidneys, readers are invited to reflect on the issue of fairness when considering initial trial inclusion and exclusion criteria. It is also not too early to think about how xenotransplantation will be implemented and how it may affect transplant systems in the United States and beyond. Indeed, an important aspect of proceeding is ensuring that research participants and future patients are able to give their freely informed consent to xenotransplantation. This is the topic of our next chapter.

References

Adams, A. B., Blumberg, E. A., Gill, J. S., Katz, E., Kawai, T., Schold, J. D., Sykes, M., Tector, A. and Sachs, D. H. 2024. Enhancing kidney transplantation and the role of xenografts: Report of a scientific workshop sponsored by the national kidney foundation. *American Journal of Kidney Diseases,* 84(1), 94–101.

Agrawal, M. and Emanuel, E. J., 2003. Ethics of phase 1 oncology studies: reexamining the arguments and data. *JAMA*, 290(8), 1075–1082.

American Medical Association. 2017. *Code of Medical Ethics*. American Medical Association.

Berry, K. N., Daniels, N. and Ladin, K. 2019. Should lack of social support prevent access to organ transplantation? *The American Journal of Bioethics*, 19(11), 13–24.

Caplan, A. L. 1985. Ethical issues raised by research involving xenografts. *JAMA*, 254(23), 3339–3343.

Cooper, D. K. and Lanza, R. P. 2000. *Xeno: The Promise of Transplanting Animal Organs into Humans*. Oxford University Press.
Cooper, D. K., Keogh, A. M., Brink, J., Corris, P. A., Klepetko, W., Pierson, R. N., Schmoeckel, M., Shirakura, R. and Stevenson, L. W. 2000. Report of the xenotransplantation advisory committee of the international society for heart and lung transplantation: The present status of xenotransplantation and its potential role in the treatment of end-stage cardiac and pulmonary diseases. *The Journal of Heart and Lung Transplantation*, 19(12), 1125–1165.
Cooper, D. K., Wijkstrom, M., Hariharan, S., Chan, J. L., Singh, A., Horvath, K., Mohiuddin, M., Cimeno, A., Barth, R. N., LaMattina, J. C. and Pierson III, R. N. 2017. Selection of patients for initial clinical trials of solid organ xenotransplantation. *Transplantation*, 101(7), 1551–1558.
Emanuel, E. J., Wendler, D. and Grady, C. 2000. What makes clinical research ethical? *JAMA*, 283(20), 2701–2711.
Food and Drug Administration. 2016. *Source Animal, Product, Preclinical, and Clinical Issues Concerning the Use of Xenotransplantation Products in Humans: Guidance for Industry*. https://www.fda.gov/media/102126/download.
Fox, R. R. C. and Swazey, J. P. 1974. *The Courage to Fail: A Social View of Organ Transplants and Dialysis*. Transaction Publishers.
Griffith, B. P., Goerlich, C. E., Singh, A. K., Rothblatt, M., Lau, C. L., Shah, A., Lorber, M., Grazioli, A., Saharia, K. K., Hong, S. N. and Joseph, S. M. 2022. Genetically modified porcine-to-human cardiac xenotransplantation. *New England Journal of Medicine*, 387(1), 35–44.
Knoll, E. and Lundberg, G. D. 1985. Informed consent and baby Fae. *JAMA*, 254(23), 3359–3360.
Liberto, H. 2014. Exploitation and the vulnerability clause. *Ethical Theory and Moral Practice*, 17, 619–629.
MacKay, D. and Saylor, K. W. 2020. Four faces of fair subject selection. *The American Journal of Bioethics*, 20(2), 5–19.
Mallapaty, S. 2024. First pig liver transplanted into a person lasts for 10 days. *Nature*, 627(8005), 710–711. https://doi.org/10.1038/d41586-024-00853-8.
Moseley, J. B., O'Malley, K., Petersen, N. J., Menke, T. J., Brody, B. A., Kuykendall, D. H., Hollingsworth, J. C., Ashton, C. M. and Wray, N. P. 2002. A controlled trial of arthroscopic surgery for osteoarthritis of the knee. *New England Journal of Medicine*, 347(2), 81–88.
National Commission for the Protection of Human Subjects of Biomedical and Behavioral Research. 1979. *The Belmont Report. Ethical Principles and Guidelines for the Protection of Human Subjects of Research*. Washington, DC: U.S. Government Printing Office, 1–20.
Nuffield Council on Bioethics. 1996. *Animal-to-Human Transplants: The Ethics of Xenotransplantation*. Nuffield Council on Bioethics.
Pascher, A., Sauer, I. M., Hammer, C., Gerlach, J. C. and Neuhaus, P. 2002. Extracorporeal liver perfusion as hepatic assist in acute liver failure: A review of world experience. *Xenotransplantation*, 9(5), 309–324.
Petrovan, S. O., Aldridge, D. C., Bartlett, H., Bladon, A. J., Booth, H., Broad, S., Broom, D. M., Burgess, N. D., Cleaveland, S., Cunningham, A. A. and Ferri, M. 2021. Post COVID-19: A solution scan of options for preventing future zoonotic epidemics. *Biological Reviews*, 96(6), 2694–2715.
Pierson III, R. N., Allan, J. S., Cooper, D. K., D'Alessandro, D. A., Fishman, J. A., Kawai, T., Lewis, G. D., Madsen, J. C., Markmann, J. F. and Riella, L. V. 2022. Expert opinion special feature: Patient selection for initial clinical trials of pig organ transplantation. *Transplantation*, 106(9), 1720–1723.

Snyder, J. 2012. Exploitations and their complications: The necessity of identifying the multiple forms of exploitation in pharmaceutical trials. *Bioethics*, 26(5), 251–258.
Tector, A. J. 2024. Xenotransplantation in humans: A reality check. *Transplantation*. https://doi.org/10.1097/TP.0000000000005223.
Wertheimer, A. 2008. Exploitation in clinical research. In *The Oxford textbook of clinical Research Ethics*. Oxford University Press, 201–210.
World Health Organization. 2008. First WHO global consultation on regulatory requirements for xenotransplantation clinical trials: Changsha, China, 19–21 November 2008. The Changsha communiqué. *Xenotransplantation*, 16, 61–63.

5 Is informed consent possible?

James Hardy (1918–2003) is a surgeon known for performing the world's first lung allotransplant at the University of Mississippi in 1963. Hardy was also determined to perform the first heart transplant, which had not been done successfully at that point, and if he could not use one from a deceased human donor, he planned to acquire one from a chimpanzee (Cooper et al., 2015). A critically ill patient presented to Hardy who, reportedly, was already in a state of decline that nowadays would exclude one as a candidate for transplantation. Nonetheless, this did not deter Hardy. As the patient did not have decision-making capacity, Hardy received consent for the surgery from the patient's sister. The consent form merely indicated that a "suitable heart" would be found for the procedure, with no mention of it possibly being a chimpanzee heart. Hardy transplanted a wild-type (meaning, it was absent of any genetic modifications) chimpanzee heart into the patient (Garnett, 1996). The patient died shortly thereafter and Hardy's reputation was tarnished. There are at least two serious ethical issues in this case—both pertaining to informed consent—to consider (i) the ethics of performing xenotransplantation, which—especially in the 1960s—was a very risky experimental surgery, on a patient unlikely to benefit and (ii) the question of whether valid consent was actually obtained from the surrogate decision-maker (Hurst, 2023).

In both the research and clinical context, informed consent is a legal requirement and an ethical imperative. In many jurisdictions today, the requirement for a researcher or clinician to obtain informed consent of the patient (or their proxy) prior to any intervention is enshrined in law. Failure to do so may constitute assault and could result in civil or criminal penalties. Yet, informed consent is also an ethical imperative that stems from respect for the autonomy of the patient. Patients have the right to determine what happens to them, and so, they should be well informed of the intervention, including potential risks, benefits, and alternatives.

DOI: 10.4324/9781003472223-6

Xenotransplantation of solid organs is currently only available in a research context. Hence, the research context—rather than the clinical—will form the majority of the discussion in this chapter, although much of the discussion applies to the clinical context as well: as noted in the previous chapter, clinical xenotransplant research is proposed to be a last ditch effort for many initial recipients and thus blurs the research/clinical distinction. In a research context, xenotransplantation presents complexities for informed consent in both adult and pediatric populations. These complexities include the risk-to-benefit ratio of the experimental therapy, the known and unknown risks that accompany receiving a cross-species transplant such as acquiring a novel infectious disease, and the potential implications brought on because of this risk. In this chapter, we discuss the development of modern-day informed consent and the challenges to informed consent brought about by xenotransplantation. Adults are the primary focus of this chapter, whereas pediatric informed consent is discussed in Chapter 6.

Informed consent

The modern-day concept of informed consent has a relatively short history. In the United States, a series of judicial decisions—beginning in 1905—played a large role in laying the foundation for patient autonomy from which informed consent derives (Bazzano et al., 2021). However, abuse of human beings under the guise of scientific research and advancement during World War II thrust research ethics to the forefront of medicine (Beauchamp and Childress, 2013, pp. 120–121). During World War II, both Nazi German and Imperial Japanese researchers committed grave acts of research misconduct on thousands of prisoners of war, criminals, and others without their consent, oftentimes resulting in their deaths (Weindling, 2008; Tsuchiya, 2008). This information came to the forefront during the Nuremberg Trials of 1945–1946 where Nazi leaders stood trial for war crimes (Persico, 1994). While consent played a role in these trials, bioethicists observe that the term *informed consent* did not appear until a decade later and was not examined in detail until the 1970s at the time of The Belmont Report (Beauchamp and Childress, 2013, pp. 120–121).

The Belmont Report is a foundational document for research ethics and the protection of human research participants in the United States. It was drafted by the National Commission for the Protection of Human Subjects and signed into federal law in 1979. The Belmont Report has three basic ethical principles: (i) respect for persons, (ii) beneficence, and (iii) justice. Respect for persons, which has subsequently evolved, for better or worse, into "respect for autonomous choice," holds that "individuals should be treated as autonomous agents" (U.S. Department of Health and Human Services, 1979). The report states,

> To respect autonomy is to give weight to autonomous persons' considered opinions and choices while refraining from obstructing their actions unless they are clearly detrimental to others. To show lack of respect for an autonomous agent is to repudiate that person's considered judgments, to deny an individual the freedom to act on those considered judgments, or to withhold information necessary to make a considered judgment, when there are no compelling reasons to do so.

The primary justification given for requiring informed consent has been the protection of the autonomous choice of the individual. Protecting autonomous choice requires the medical provider or researcher to meet certain standards for informed consent to be valid.

Standards for informed consent

Standards for informed consent differ by jurisdiction. We write largely from a US perspective due to this being the location where clinical solid organ xenotransplantation will most likely begin. The Belmont Report, foundational to research ethics in the United States, emphasizes that persons should have the opportunity to decide what shall or shall not happen to them in a medical context and that this opportunity is only provided when certain standards are met. Three standards are provided in The Belmont Report: (i) information, (ii) comprehension, and (iii) voluntariness. Taking these in turn, adequate information must be disclosed to persons, which generally include: "the research procedure, their purposes, risks and anticipated benefits, alternative procedures (where therapy is involved), and a statement offering the subject the opportunity to ask questions and to withdraw at any time from the research" (U.S. Department of Health and Human Services, 1979). Second, researchers have an obligation to ensure that research participants adequately comprehend the information that is disclosed. Third, participation in research is valid only if consent is voluntarily given. This means that the participant must be free of coercion (which typically includes a threat of harm) and undue influence (an offer of excessive reward or benefit). Unjustifiable pressure that falls under the category of undue influence compromises voluntariness, though The Belmont Report articulates the difficulty with this element, as "it is impossible to state precisely where justifiable persuasion ends and undue influence begins." Examples of undue influence provided in The Belmont Report include "manipulating a person's choice through the controlling influence of a close relative and threatening to withdraw health services to which an individual would otherwise be entitled," though this second example seems more closely aligned with coercion as it includes an overt threat of harm.

The Code of Federal Regulations, which succeeded The Belmont Report, sets standards for informed consent in the United States. From a regulatory perspective, this falls under the purview of the United States Department of Health and Human Services. Based upon the United States federal requirements for informed consent, several ethical issues must be considered for xenotransplantation, including—but not limited to—risks of undue inducement (another term for "influence"), the potential problem of being unable to withdraw from biosurveillance, accurate disclosure of reasonably foreseeable risks or discomforts or limits on personal freedom, the media attention that may ensue, and appropriate disclosure about potential benefits. These aspects will be discussed in the following section.

Challenges to informed consent in xenotransplantation

As seen in the preceding sections, informed consent is an essential component of medical research. However, xenotransplantation presents challenges to informed consent in the context of clinical trials and may continue to do so if xenotransplantation becomes a clinical option. In 1998, Harold Vanderpool remarked that the informed consent process from patients in xenotransplantation will be exceedingly challenging, possibly surpassing those of phase 1 cancer trials. Vanderpool bases his conclusion on the desperate condition of the individuals who will likely be recruited for these clinical trials—patients with organ failure who are unlikely to receive an allotransplant soon, if at all. Desperation, then, raises the question of whether a person can give consent freely: does desperation compromise a person's voluntariness in deciding to participate in a clinical trial? This worry arises in allotransplantation as well. Paul Ramsey, a founder of modern bioethics, voiced this concern in regard to the burgeoning field of heart transplantation in his day (2002, p. 223). Voluntariness is recognized to be essential to informed consent (Beauchamp and Childress, 2013, p. 124). It is understood that consent must be free from coercion (which typically includes a threat to someone's rights or threat of harm), exploitation (taking unfair advantage), and undue inducement.

Undue inducement

Undue inducement, or what is sometimes termed undue influence, is mainly in view here for xenotransplantation. A common example of undue inducement is paying research participants for their participation in a clinical trial. Paying persons to enroll in a clinical trial is standard practice.

However, there is a worry that paying people too much—whatever that subjective amount might be—runs the risk of unduly inducing a person to participate. Ezekiel Emanuel (2005) has further defined undue inducement, arguing that it includes an offer to provide a reward that is excessive and results in the potential research participant exhibiting poor judgment and possibly making a choice they would not have otherwise made. That is, the participant is made an offer they cannot reasonably refuse but could do so if there were no "undue" inducement. Emanuel (2005) further defines four distinct elements that must be present for undue inducement to occur: (i) a good (something that is desirable) must be offered to the potential research participant; (ii) the good that is offered is excessive (i.e., it is not easily refused); (iii) the offer of an excessive good leads the potential research participant to exercise poor judgment (i.e., their judgment is clouded by the excessive good); and (iv) there is risk of serious harm occurring.

To be clear, mere inducement is not the worry here. Persons are induced to a behavior commonly without any ethical quandary. You are induced to perform well at your job in order to have job security and perhaps receive a promotion or salary increase. A clothing retailer wants to increase business and has a sale on clothing to induce people to frequent their store. Such inducements to modify behavior are common and without much ethical quandary. We do it all the time.

In the case of xenotransplantation clinical trials, what might constitute an *undue* inducement? Let's recall—an inducement may be *undue* if it is somehow excessive and results in the potential research participant exhibiting poor judgment and possibly making a choice they would not have otherwise. Clinical trials of xenotransplantation will likely involve persons who have run out of other good options (see Chapter 4). For kidney xenotransplantation clinical trials, persons will likely already be on some form of renal replacement therapy and be listed far down on the kidney waitlist, making it unlikely they would receive one in time. In the United States, dialysis has been the standard treatment for patients with kidney failure since the 1970s. Hemodialysis is not an enjoyable experience. The majority of hemodialysis in the United States is performed in outpatient centers where patients must travel approximately three times per week, with each session lasting three to four hours. Not only is this a time-consuming treatment, many patients also leave exhausted; hence, it does not provide a great quality of life. Certainly, it is not difficult to see how a patient without much hope of receiving an allotransplant could be induced to undergo xenotransplantation. The question is whether or not the lack of other good options and the offer of an experimental porcine kidney is excessive, resulting in the patient exhibiting poor judgment, and thereby compromising their voluntary informed consent.

Xenotransplantation clinical trials are not the only type of novel medical research that must consider undue inducements. An analogy can be made to early phase oncology clinical trials. Let us suppose a patient has pancreatic cancer and standard therapy of radiation and chemotherapy has been tried without success. Surgery is not an option and so the patient's only recourse is now palliative care. However, the oncologist approaches the patient with the potential for a new chemotherapy clinical trial option from investigators at a neighboring medical center. This is a phase 1 clinical trial that seeks to enroll about 15 patients with pancreatic cancer, and it is still unknown how safe or effective the investigative medicine is in humans. The patient fits the inclusion criteria and is hopeful this will offer a cure for her cancer. While there may be a problem in her reasoning due to therapeutic misconception, let us set that aside to determine whether she is unduly induced to participate in the clinical trial. If we utilize Emanuel's criteria above, we can see that this is at least a potential good (criterion i) and that there is the possibility of excessive harm (criterion iv) occurring due to this being an experimental therapy. Yet it is not altogether certain that the offered good is excessive (criterion ii) and the offer of an excessive good leads the potential research participant to exercise poor judgment (criterion iii). That is, the criteria are not met. Hence, the desirability of the "solution" proffered by an early phase clinical trial (or xenotransplant for that matter) does not introduce any obvious distorting effects that lead to poor judgment. There is inherent ambiguity in whether an offer is excessive and presents *unreasonable* risk to a research participant. This leads Emanuel to conclude that undue inducement cannot occur if a clinical research study is otherwise ethically conducted because there is no risk of excessive risks—risks that rise above what a reasonable person would assume.

We tend to agree with Emanuel's conclusion and do not believe that the first-in-human xenotransplantation clinical trials are altogether different, from a risk standpoint and in terms of presenting a "last-ditch lifesaving effort" to a patient, from an early phase oncology clinical trial. The problems facing phase 1 oncology trials are well-known insofar as the decisions of participants to enroll are often beset by therapeutic misconception and unrealistic optimism. The strategies used in the phase 1 context to address these problems can be used to address them in xenotransplantation. Potential research participants should be informed of their options and that their end-of-life care might be compromised by the burdens associated with the study and that they might have more quantity and quality of life by not participating. Yet, if the patient desires to assume the risk to pursue the potential benefits of a xenograft, however unlikely, then we see no objection. As long as the potential research participant has received an opportunity for thorough informed consent on the known and unknown risks of xenotransplantation, as well as the potential that the patient will not receive any direct benefit, participating seems ethical.

Potential inability to withdraw from the clinical trial

When a pathogenic microorganism is transmitted from a nonhuman animal to a human, it is described as a zoonosis or zoonotic disease or infection. The pathogenic microorganism may be a virus, bacteria, parasite, or fungi. Such infections are common occurrences and can beget novel diseases in humans. In recent years, this has been seen with Ebola, avian flu, swine flu, and perhaps SARS-CoV-2. The potential for a pathogenic microorganism to be transmitted from the porcine graft to the human recipient, and then possibly to others, has spurred several organizations and individual commentators to argue that xenotransplantation recipients should undergo regular biosurveillance (e.g., graft biopsy, imaging, and blood tests) for xenozoonotic infection. In essence, this would be a mandatory aspect of the clinical trial from which the research participant could never withdraw (see Chapter 3), which is contrary to the United States' research guidelines (21 CFR 50.25) that mandate the "subject may discontinue participation at any time without penalty or loss of benefits to which the subject is otherwise entitled." This is perhaps the most serious concern facing the consent process in xenotransplant research.

Due to the supposed need for a xenotransplantation recipient to be monitored for infectious disease for the remainder of their life, a mechanism for trying to enforce this has been considered, with some proposing a contract of sorts. In the realm of psychiatry, a psychiatric advance directive, also known as a "Ulysses" contract, is a document that a psychiatric patient, during a state of lucidity and in consultation with a medical professional, fills out and agrees to regarding their care at some future point when experiencing a state in which their decision-making capacity is impaired. The name "Ulysses contract" is derived from the ancient Greek epic poem by Homer, *The Odyssey*. In the story, the protagonist Odysseus (Latinized as "Ulysses") has heard tales of the legendary Sirens—mythological creatures with beautiful, alluring voices. Their beautiful voices were known to entrance captains of passing ships, luring them off their course and toward the rocks where the ship would break apart and all aboard perish.

On a sea voyage with his crew, Ulysses was on course to sail past the Sirens. He harbored a curiosity to listen to their enchanting singing, yet his desire to survive and recount the experience was equally strong. The goddess Circe advised Ulysses on how to achieve this delicate balance:

> Therefore pass these Sirens by, and stop your men's ears with wax that none of them may hear; but if you like you can listen yourself, for you may get the men to bind you as you stand upright on a cross-piece half way up the mast, and they must lash the rope's ends to the mast itself, that you may have the pleasure of listening. If you beg and pray the men to unloose you, then they must bind you faster.
>
> (Homer)

Hence, prior to losing "decision-making capacity" (due to the entrancement of the Sirens), Ulysses ordered his fellow sailors to only heed his commands to untie him once they had passed out of earshot of the Sirens.

Some commentators have attempted to appropriate Ulysses contracts to xenotransplantation, arguing that they could be used to mandate post-transplant monitoring. For instance, in 1999, Daar proposed that xenograft recipients could sign a Ulysses contract that would then compel them to abide by the monitoring requirements. Several commentators have since similarly argued for the use of Ulysses contracts as it would supposedly provide a mechanism to circumvent federal regulations that a research participant can withdraw from any aspect of a clinical trial (including posttransplant monitoring) at any point. However, the applicability of Ulysses contracts to xenotransplantation is questionable because (i) the end goal of the advance directive may not be congruous to the xenotransplantation clinical context, (ii) enforcing Ulysses contracts in xenotransplantation is impractical, and (iii) the ethical and regulatory hurdles that such enforcement would require are difficult to justify (Hurst et al., 2024a). Specifically, in a Ulysses contract for psychiatric reasons—the context they are most often used—the patient does not have decision-making capacity when the contract goes into effect. Yet, post-xenotransplantation, it is not expected that a patient would have *compromised* decision-making capacity. Hence, there seems to be no clinical indication for a Ulysses contract in xenotransplantation clinical trials.

There is a conflict, therefore, between public health considerations and respect for autonomy. On the one hand, given the concerns of zoonotic infection transmission, researchers suggest suspending recipients' right to withdraw; on the other hand, there are ethical and legal challenges to such a proposal that protect the right to withdraw. If informed consent were changed in this context such that one could sign away the right to withdraw, might it be changed in others? There does not seem to be any guarantee that such proposed revisions to the consent process are limited to xenotransplantation.

Potential restrictions on reproductive rights

In 2001, the United States Public Health Service issued guidelines on infectious diseases transmission in xenotransplantation. The guidelines cautioned about the possible risk of zoonotic disease transmission and the reproductive risk this may entail:

> 2.5.12. Xenotransplantation product recipients who may wish to consider reproduction in the future should be aware that a potential risk of transmission of xenogeneic infectious agents not only to their partner but also to their offspring during conception, embryonic/fetal development and/or breast-feeding cannot be excluded.

At a minimum, dependent on the xenograft recipient being of reproductive age, this potential risk should be included in the informed consent process. However, a separate issue is that even should participants be counseled not to reproduce and they agree to this, enforcement seems difficult if not impossible (Padilla et al., 2022). Asking potential participants to refrain altogether from sexual intercourse would likely be met with resistance and, again, difficult to enforce. The United Kingdom's Nuffield Council on Bioethics (1996) stated that asking participants to refrain from having children due to the risk of a virus transmitted from parent to child would be unacceptable. Yet, David Cooper (2023) published a draft of an informed consent document for a xenotransplantation clinical trial which states that men capable of fathering a child should either abstain from sex or take measures to prevent pregnancy. Similarly, he sees it as an exclusionary criterion to enroll women of childbearing age unless they have undergone hysterectomy or tubal ligation. Given these stipulations, the risk of potential zoonotic transmission via sexual activity to partners as well as offspring must be included in the informed consent process. However, requiring participants to refrain from sexual activity or attempting to reproduce is unenforceable. It is also worth questioning that if all premenopausal women must have received hysterectomy or tubal ligation prior to xenotransplantation, should men have received a vasectomy? If the risk of possible xenozoonotic transmission is high enough to warrant such interventions for preventing pregnancy in women, should the same standard not also be applied to men? After all, men retain the ability to reproduce for a longer period of time than women do.

Isolation and quarantine

If a xenotransplantation recipient or their contact(s) begins exhibiting symptoms of unexplained illness, or if zoonotic infection is detected following biosurveillance measures, this may be the cause for isolation or quarantine. Isolation typically refers to segregating sick persons from those who are not sick. Whereas, quarantine segregates and restricts the movement of a person who has had some contact with a communicable disease to see if sickness develops. In both cases, the objective is the same: to prevent any further transmission of disease. Insofar as the risk of xenozoonotic disease and subsequent quarantine is known in xenotransplantation, it should be disclosed in the informed consent process.

Forced quarantine has raised the concern of violating an individual's freedom by pitting the individual's freedom against the welfare of the public (Padilla et al., 2022). However, legal authority exists in the United States at the federal and state levels to quarantine individuals with the use of police powers if necessary. The Commerce Clause of the United States Constitution is cited as the origin of the authority for isolation and quarantine procedures, and the Public Health Service Act authorizes the United States Secretary of Health and Human Services to prevent the transmission of communicable diseases. This authority is then delegated to the United States Centers for Disease Control and Prevention (CDC) for carrying out the daily functions. The CDC has the authority to detain, medically examine, and quarantine an individual under Title 42 Code of Federal Regulations parts 70 and 71. Individual states have police powers to protect persons within their jurisdictions and can enforce isolation and quarantine procedures. Hence, isolation and quarantine could be necessary post-xenotransplantation for the xenograft recipient and, possibly, their contacts.

An issue is whether isolation and quarantine can be enforced on a xenograft recipient who displays no symptoms of illness. If a recipient fails to comply with monitoring protocols but is otherwise healthy, should that person be quarantined? United States law allows public health officials to isolate or quarantine individuals who are symptomatic or may become symptomatic because of disease contact. There is no precedent in United States law to quarantine someone who is asymptomatic and who has not had close contact with someone who was sick.

Media attention

The first (January 2022) and second (September 2023) cardiac xenotransplantations under FDA expanded access attracted significant media attention. Some of this was generated by the hospital—The University of Maryland Medical Center—which issued press releases. Likely, this was arranged with the patient and family during the informed consent conversations prior to surgery, as the patient's names were released in the press releases which would require the patient's explicit consent. The hospital system, it would seem, desired the media attention as the press release speaks of it being a "historic first," "first-of-its-kind," and a "breakthrough surgery" (Kotz, 2022).

A team of researchers at the University of Maryland Medical Center has recommended that potential xenograft recipients be educated on the possibility of media exposure for as long as xenotransplantation is newsworthy—which would seem to be at least during expanded access studies and clinical trials (Cherukuru, 2023). Hospitals have a fiduciary duty to ensure that sensitive health information, which is further protected by law, is kept confidential. In the case of medical breakthroughs, some level of media attention is warranted on the breakthrough itself, yet the media will also be curious about the individuals who participated. It is worth noting that predicating inclusion in a xenotransplantation clinical trial on whether a patient agrees for their name to be released and media stories written about them seems disingenuous. The priority of medicine is care for patients, not public recognition. Nevertheless, if a patient agrees that their name can be released as having received a xenograft, then they should be educated on what such exposure might entail for them and their families, and perhaps offered "public relations" support as a post-xenotransplantation obligation as needed.

Risk-benefit ratio

Adequately describing the possible risks and benefits of any novel therapy can be difficult. Indeed, some uncertainty in the risks and benefits of clinical trials is inherent, especially early on. Relevant preclinical evidence is used to determine potential risks and benefits in human studies, although some criticize the translatability of preclinical data (Kimmelman and Henderson, 2016). For xenotransplantation clinical trials, this would include pig-to-baboon studies as well as the decedent studies involving "brain-dead" recipients and expanded access cases that have been conducted. Even with this evidence, it seems that first-in-human xenotransplantation clinical trial participants will assume a high level of risk with tenuous benefits.

An aspirational if not accepted viewpoint is that clinical research can be ethically justified only if it has a favorable risk-benefit ratio—a requirement that embodies fundamental bioethical principles of beneficence and nonmaleficence (Emanuel et al., 2000). Phase 1 clinical trials, though, are primarily safety studies in which researchers investigate toxicity levels in which a positive effect could only happen late in the trial when between 80 and 120% of the adequate dosage is administered, and this relatively unknown chance of benefit is based of inferential conjectures drawn from animal studies and perhaps other data from similar classes of medicines. Hence, researchers try to take precautions by giving dosages at the lower ends of what they believe the therapeutic range to be, yet there is still much unknown about potential side effects. Similarly, in early xenotransplantation clinical trials, it seems likely that researchers cannot expect patients to live long (at this point) with a xenograft. The cardiac xenotransplants that took place in 2022 and 2023 evidence this. How can such a study have a favorable risk-benefit ratio?

Emanuel and colleagues (2000), as well as others such as Charles Weijer (1999), advocate that in circumstances when no or very little potential benefits to individual research participants exists, or when the risks of the proposed intervention clearly outweighs any potential benefit to the individual, the potential for societal benefit should also hold weight. Emanuel admits that no settled framework exists for how to weigh potential societal benefit, potential individual benefit, and individual risk. Yet, Emanuel holds that if potential individual risk outweighs any potential benefit to the individual or society, the research is not ethically justifiable. If Emanuel's framework is accepted for xenotransplantation, then early xenotransplantation clinical trials could be ethically justified only if there is potential for societal benefit, as the potential for individual benefit is currently rather marginal.

There is a tension here. People are being asked to undergo a risky procedure largely for the benefit of others, and we should wonder if we are reaching an upper limit to what we can ask people to sacrifice. Some, like MacKay and Saylor (2020), believe that "in the context of the cooperative project it is unfair to ask some individuals to bear excessive burdens for the purposes of benefiting others." To be sure, as discussed in Chapter 4, acceptable initial research participants will be those for whom therapeutic alternatives are unavailable: it is *these* people who are being asked to risk it all for the benefit of society. Medical research, as Kong (2005) observes, is a social undertaking that is pursued for the benefit of society, and it thus reflects social values with those engaged in such research operating as societal agents. Current guidelines suggest that xenotransplantation research is ethically defensible on the seriously ill but not defensible on the non-seriously ill, and it could be inferred that this manifests an assumption that the seriously ill have less value than the non-seriously ill. As Kong (2005) explains,

> it does not follow that research which is unethical in a fit young adult is suddenly ethical because the patient is dying. . . . The dying have as much right to not be harmed or used as the healthy.

Nonadherence to treatment posttransplant

Nonadherence to the treatment plan posttransplant is a significant risk factor for rejection in allograft recipients, and it is likely to be similar in xenograft recipients. Recipient selection therefore should favor individuals that display evidence of resiliency, especially since xenotransplantation follow-up poses unique challenges (Reese et al., 2023; Nellore et al., 2022). Receiving a new organ requires a lot of care and the associated risks of zoonosis and possible stigma associated with having a nonhuman organ may be burdensome for individuals, thereby making it easier to withdraw from follow-up; this is especially true given that follow-up will be invasive and recipients may feel otherwise healthy. To mitigate these burdens, a standard of psychological and social support should be met by would-be recipients. For instance, previous medical nonadherence, untreated depression, or lack of psychological and social support should disqualify someone from participating in a xenotransplant trial (Reese et al., 2023; Silverman and Odonkor, 2022). There is room for debate about how to assess for resiliency, but the standard should be high given what is being asked of would-be recipients and the challenges they will face. Because the high standard may risk increasing exclusions, putting the trial at risk, and exacerbating injustices, putting the welfare of those wanting to participate at stake, strategies for ameliorating a lack of psychosocial support ought to be explored.

Recommendations

As can be seen from the above, the challenges to informed consent in xenotransplantation are many, though they are not insurmountable. Among the immense amount of information that must be presented to potential xenotransplant recipients (or a surrogate decision-maker), researchers should explain in comprehensible detail the risk of zoonosis, the importance of follow-up, and what follow-up will include: a lifelong commitment to aggressive immunosuppression therapy conjoined with invasive biosurveillance. Foreseeable risks and discomforts should be explained in an easy-to-comprehend manner (Caplan and Parent, 2022). To facilitate comprehension, would-be recipients might be given a mandatory decision period of one to two days before being allowed to sign the informed consent document. As Padilla and colleagues explain, "IRBs could consider a 2-step signature consent process (or 'cooling off period') in which the same [informed consent] document is provided on 2 separate occasions" (2022, p. 1758). Second, prior to signing, would-be recipients should be given an assessment test that evaluates their understanding of the risk of zoonosis and the importance of follow-up. If the recipients fail the assessment test, they should not be allowed to enroll at that point, and more information should be provided.

It is also important to address during the informed consent stage the financial burdens associated with lifelong biosurveillance. To receive a xenograft, a patient will have to agree to travel to and from a medical center, which may incur significant cost. Due to the availability (or lack thereof) of biosurveillance in a certain area, there may be restrictions on where the recipient can live (e.g., living abroad for work may be strongly discouraged). Hence, would-be recipients should be informed of possible long-term financial costs of receiving a xenograft, and every effort should be made by the medical team and sponsoring company to mitigate financial burden (Reese et al., 2023).

Given the importance of third-party monitoring, and the fact that a recipient's loved ones may be tracked for possible zoonosis, the recipient's loved ones should, ideally, assent to lifelong monitoring and follow-up (Silverman and Odonkor, 2022). U.S. Department of Health and Human Services (2001) guidelines posit the inclusion of a "documented procedure to inform the recipient of the responsibility to educate his/her close contacts" of the risks of zoonosis (2.5.4); however, a better approach would be that initial screening include close contacts being informed and educated by the medical team (Hurst et al., 2024b). Receiving a xenograft could place close contacts in danger, and as such, close contacts should be informed of what the loved one is undergoing. Since close contacts are not themselves undergoing the medical procedure, they do not need to give informed consent; rather, assent to possible biosurveillance should be enough.

Conclusion

As xenotransplantation continues to advance, recommendations for best practice in regard to informed consent will need to be continually analyzed. This should be a multidisciplinary effort not only consisting of xenotransplantation investigators and physicians but also including ethicists, social workers, psychiatrists/psychologists, clergy persons, and other professionals. The ethical and legal challenges to thoughtfully securing the informed consent of adult patients are worth serious discussion. Of course, the challenges—monitoring, quarantine/isolation, potential reproductive restrictions, and the like—are further complicated when considering children, especially younger children, who rely on their surrogate, usually parents, to make medical decisions on their behalf, and who stand to benefit from xenotransplantation. We turn to the unique challenges posed by pediatric xenotransplantation next.

References

Bazzano, L. A., Durant, J. and Brantley, P. R. 2021. A modern history of informed consent and the role of key information. *Ochsner Journal*, 21, 81–85.

Beauchamp, T. L. and Childress, J. F. 2013. *Principles of Biomedical Ethics*. 7th edition. Oxford: Oxford University Press.

Caplan, A. and Parent, B. 2022. Ethics and the emerging use of pig organs for xenotransplantation. *The Journal of Heart and Lung Transplantation*, 41, 1204–1206.

Cherukuru, N., Athanasiadi, A., Lemalefant, R., Mancini, D., Forrester, A., Glovinsky, D., Miski, P., Harrison-Restelli, C. and Robinson, C. 2023. Lessons learned from the first cardiac xenotransplant in a consciously consenting human: Psychiatric considerations and the impact of media exposure. *Xenotransplantation*, 30, e12830.

Cooper, D. K. C. 2023. Patient informed consent for a clinical trial of gene-edited pig kidney transplantation: A representative consent form. *Xenotransplantation*, 30, e12790.

Cooper, D. K. C., Ekser, B. and Tector, A. J. 2015. A brief history of clinical xenotransplantation. *International Journal of Surgery*, 23, 205–210.

Daar, A. S. 1999. Xenotransplantation: Informed consent/contract and patient surveillance. *Biomedical Ethics*, 4, 87–91.

Emanuel, E. J. 2005. Undue inducement: Nonsense on stilts? *American Journal of Bioethics*, 5, 9–13; discussion W8–11, W17.

Emanuel, E. J., Wendler, D. and Grady, C. 2000. What makes clinical research ethical? *JAMA*, 283, 2701–2711.
Garnett, R. W. 1996. Why informed consent? Human experimentation and the ethics of autonomy. *Catholic Lawyer*, 36, 455–512.
Homer. *The Odyssey*. Translated by Samuel Butler. Project Gutenberg. https://www.gutenberg.org/cache/epub/1727/pg1727-images.html (Accessed: 4 September 2024).
Hurst, D. J. 2023. Xenotransplantation and informed consent. In Hurst, D. J., Padilla, L. and Paris, W. D. (eds.), *Xenotransplantation: Ethical, Regulatory, and Social Aspects*. Cham: Springer International Publishing.
Hurst, D. J., Padilla, L., Schiff, T. and Parent, B. 2024a. Revisiting the use of Ulysses contracts in xenotransplantation. *Transplantation*, 108, 369–373.
Hurst, D. J., Padilla, L., Rodger, D., Schiff, T. and Cooper, D. K. C. 2024b. Close contacts of xenograft recipients: Ethical considerations due to risk of xenozoonosis. *Xenotransplantation*, 31, e12847.
Kimmelman, J. and Henderson, V. 2016. Assessing risk/benefit for trials using preclinical evidence: A proposal. *Journal of Medical Ethics*, 42, 50–53.
Kong, W. M. 2005. Legitimate requests and indecent proposals: Matters of justice in the ethical assessment of phase I trials involving competent patients. *Journal of Medical Ethics*, 31, 205–208.
Kotz, D. 2022. *UM Medicine Performs Historic Xenotransplantation*. https://www.umaryland.edu/news/archived-news/january-2022/um-medicine-performs-historic-xenotransplantation.php (Accessed: 12 January 2022).
MacKay, D. and Saylor, K. W. 2020. Four faces of fair subject selection. *American Journal of Bioethics*, 20, 5–19.
Nellore, A., Walker, J., Kahn, M. J. and Fishman, J. A. 2022. Moving xenotransplantation from bench to bedside: Managing infectious risk. *Transplant Infectious Disease*, 24, e13909.
Nuffield Council on Bioethics 1996. *Animal-to-Human Transplants: The Ethics of Xenotransplantation*. London: Nuffield Council on Bioethics.
Padilla, L. A., Hurst, D., Maxwell, K., Gawlowicz, K., Paris, W., Cleveland, D. and Cooper, D. K. C. 2022. Informed consent for potential recipients of pig kidney xenotransplantation in the United States. *Transplantation*, 106, 1754–1762.
Persico, J. E. 1994. *Nuremberg: Infamy on Trial*. New York: Penguin.
Ramsey, P. 2002. *The Patient as Person: Explorations in Medical Ethics*. 2nd edition. New Haven: Yale University Press.
Reese, P. P., Gelb, B. E. and Parent, B. 2023. Unique problems for the design of the first trials of transplanting porcine kidneys into humans. *Kidney International*, 103, 239–242.
Silverman, H. and Odonkor, P. N. 2022. Reevaluating the ethical issues in porcine-to-human heart xenotransplantation. *Hastings Center Report*, 52, 32–42.
Tsuchiya, T. 2008. The imperial Japanese experiments in China. In Emanuel, E. J., Grady, C. A., Crouch, R. A., Lie, R. K., Miller, F. G. and Wendler, D. (eds.), *The Oxford Textbook of Clinical Research Ethics*. Oxford: Oxford University Press.
U.S. Department of Health and Human Services. 1979. *The Belmont Report*. https://www.hhs.gov/ohrp/regulations-and-policy/belmont-report/index.html (Accessed: 5 March 2024).
U.S. Department of Health and Human Services. 2001. *PHS Guideline on Infectious Disease Issues in Xenotransplantation*. https://www.fda.gov/media/73803/download (Accessed: 1 March 2024).

Vanderpool, H. Y. 1998. Critical ethical issues in clinical trials with xenotransplants. *The Lancet*, 351, 1347–1350.
Weijer, C. 1999. Thinking clearly about research risk: Implications of the work of Benjamin Freedman. *IRB: Ethics & Human Research*, 21, 1–5.
Weindling, P. 2008. The Nazi medical experiments. In Emanuel, E. J., Grady, C. A., Crouch, R. A., Lie, R. K., Miller, F. G. and Wendler, D. (eds.), *The Oxford Textbook of Clinical Research Ethics*. Oxford: Oxford University Press.

6 What interests are at stake for children?

In October 1984, Stephanie Fae Beauclair—known as "Baby Fae"—became the world's first infant to receive a cardiac xenotransplantation from a baboon. Baby Fae was born with hypoplastic left heart syndrome (HLHS), a rare congenital heart defect in which the left side of the heart is underdeveloped and does not pump well. While more effective surgical interventions aside from transplantation are available for some infants today, the cardiac surgery field was much different in 1984. Allotransplantation and the so-called "Norwood procedure" were the most viable alternatives at the time, yet both of these were judged to be unfavorable, a judgment that would become a source of controversy (Knoll and Lundberg, 1985; Jonasson and Hardy, 1985). Even today, finding a suitable donor, especially a pediatric donor, is exceedingly difficult, meaning that many children like Baby Fae will die. Thus, the situation faced by Baby Fae's parents was dire, and remains so today for a certain population of children.

Dr. Leonard Bailey, a pediatric cardiac surgeon at Loma Linda University Hospital in California, was interested in xenotransplantation, having performed sheep-to-goat xenotransplantation studies. Bailey became interested in using immature baboons as sources of hearts for neonates with HLHS. Loma Linda purchased a panel of infant baboons for Bailey to begin studying for infectious diseases and compatibility with human neonates (Bailey, 2011). He eventually presented the data to the Loma Linda University institutional review board (IRB) in a research protocol titled, "Orthotopic Cardiac Xenotransplantation in the Newborn with Diminutive Left Heart Syndrome" (Dommel, 1986). The IRB approved the protocol with the stipulation that the protocol also be approved by various other committees at the hospital, such as the ethics committee and the departments of surgery and pediatrics. After consulting with experts outside Loma Linda University, final approval to proceed with baboon-to-human cardiac xenotransplantation in an infant, should one present and the parent(s)/guardian(s) consent, was received. Shortly thereafter, Baby Fae would be

DOI: 10.4324/9781003472223-7

born and transferred to Loma Linda University Hospital. Specialists were consulted on Baby Fae's condition and the lack of options available to her. Dr. Bailey then presented the experimental option of receiving a baboon heart. After much discussion over subsequent days, both parents agreed. Despite ABO-incompatibility between Baby Fae and baboon, the heart functioned well after transplant for two weeks. However, sadly, the heart underwent rejection and Baby Fae would die on postoperative day 20.

While the Baby Fae case occurred over four decades ago now, the cardiac allotransplant waitlist continues to suffer from lack of supply. While certain alternatives to allotransplant have been developed, these do not always work well in small children. Several advancements have been made that make beginning cardiac xenotransplantation clinical trials in pediatrics possible. However, certain ethical issues in this vulnerable population must be carefully considered. Children are unable to decide for themselves, and as discussed in the previous chapter, informed consent to xenotransplantation carries potentially significant lifelong consequences, not to mention serious risk and (current) uncertain benefit. This chapter will first provide a background on the need for pediatric organs and how xenotransplantation may fill this shortage of organs. We will then discuss specific ethical issues attendant with pediatric xenotransplantation, including whether pediatric clinical trials should precede adult trials, issues regarding informed consent, psychosocial considerations, and informed consent and the need for lifelong monitoring.

The need for pediatric organs

We have discussed in various places the dearth of available organs *vis-à-vis* their need in the adult population. Healthy organs for transplantation are in short supply for the pediatric population—especially for younger children. In the United States alone, nearly 2,000 children are on the transplant waitlist (Health Resources and Services Administration), with about one-quarter of these children being under the age of 5. Size-matching of organs is critical in this population, as children often have better outcomes if they receive a similar child-sized organ. Hence, a child is typically not a candidate for an adult organ. While livers, hearts, and kidneys are all needed by children on the waitlist due to conditions that can occur during gestation, in infancy, or later in childhood, in this chapter, we are primarily focused on pediatric cardiac xenotransplantation, as it is most likely the soonest to occur in children.

In 2021, over 700 patients were added to the pediatric heart transplant waitlist, which brought the total to over 1,000 (U.S. Department of Health and Human Services, 2021). For younger children, those under one year of age, in need of a solid organ transplant, the majority are waiting for a heart. While this number is lower than the adult waitlist, this population group—neonates and infants—faces high waitlist mortality as there are few options for mechanical circulatory support. Around half of the infants on mechanical circulatory support while they await a heart transplant die within six months (Blume et al., 2018). Transplantation is their best hope, yet the supply of healthy pediatric hearts for transplant is limited. Therefore, it is no wonder that alternatives to allotransplantation are being sought. In fact, one biotechnology company announced plans in mid-2023 that they wanted to begin pediatric cardiac xenotransplantation the following year (Hamzelou, 2023). As of this writing, those trials have yet to begin.

Proceeding with pediatric xenotransplantation clinical trials

In recent years, due to the advancements made in xenotransplantation and the great need for alternatives to allotransplantation for pediatric patients, there has been greater discussion about initiating cardiac xenotransplantation clinical trials in this population before adults. The argument can be presented like this:

1) Researchers and clinicians have a moral responsibility to try to meet the need for pediatric hearts.
2) Xenotransplantation may help meet the need.
3) In order for xenotransplantation to help meet the need, xenotransplant research in pediatric patients needs to be conducted.
4) Researchers and clinicians have a moral responsibility to try to meet the need for pediatric hearts by conducting xenotransplant research in pediatric patients.

This argument has recently been proposed at length in certain publications (Hurst and Padilla, 2024; Hurst et al., 2024). The argument is similar to that offered in defense of xenotransplant research discussed in Chapters 1 and 2: xenotransplantation holds promise to save many human lives and alternatives are currently implausible or more ethically controversial.

However, there is an opposing viewpoint that is supported by certain internationally recognized ethics guidelines. The opposing viewpoint holds that experimental therapies should first be conducted in adults before children. For example, the Nuffield Council says,

> In the case of early xenotransplantation trials, the main problem is likely to be overcoming organ rejection, which will be a problem in both adults and children. Thus, it would be difficult to justify the involvement of children in major and risky trials as recipients of heart xenografts, for

> example, before some of the uncertainties have been eliminated in trials involving adults.
>
> (Nairne and Nuffield Council on Bioethics, 1996, p. 90)

Since those uncertainties have not been resolved, children ought not to be involved. The underlying rationale is that children are vulnerable and worthy of special protections. Children—particularly infants—lack agency and autonomy; they are unable to decide for themselves. For this reason, they are especially vulnerable to exploitation, coercion, and abuse. To conduct risky research on children rather than on adults who are capable of informed consent (Chapter 5) is controversial; parental consent may be undermined by therapeutic misconception on account of the risk of death; the burdens of long-term surveillance may apply for a much longer amount of time than for an older adult.

The Belmont Report, a seminal research ethics guideline in the United States that forms the bedrock for the protection of human research subjects, states in its section on selection of research subjects the following:

> Social justice requires that distinction be drawn between classes of subjects that ought, and ought not, to participate in any particular kind of research, based on the ability of members of that class to bear burdens and on the appropriateness of placing further burdens on already burdened persons. Thus, it can be considered a matter of social justice that there is an order of preference in the selection of classes of subjects (e.g., adults before children) and that some classes of potential subjects (e.g., the institutionalized mentally infirm or prisoners) may be involved as research subjects, if at all, only on certain conditions.
>
> (U.S. Department of Health and Human Services, 1979)

The Council for International Organizations of Medical Sciences' (CIOMS's) "International Guidelines for Health-related Research Involving Humans," a set of research ethics guidelines prepared in collaboration with the World Health Organization, is more specific. CIOMS bifurcates its commentary into two categories: research with no potential for benefit and research that does have potential to benefit children. For research that has no potential individual benefits for the participants, CIOMS (2016) holds that the intervention must first be studied in adults when possible. If the data cannot be studied without the participation of children, then "risks must be minimized and no more than minimal." However, if research does have the potential for benefit, CIOMS does not require that the research first be conducted in adults:

> This prospect [for potential individual benefit] is sufficient to justify the risks associated with the interventions and procedures, provided that

> the cumulative risk of all study interventions and procedures that do not have a prospect of potential individual benefit are no more than minimal. If research meets these conditions but the cumulative risk of all study interventions and procedures that do not have a prospect of potential individual benefit is only a minor increment above minimal risk, then research ethics committees must be convinced that the research is of special relevance to children or adolescents and could not be carried out equally well in an adult population.
>
> (CIOMS, 2016)

Yet, with first-in-human pediatric cardiac xenotransplantation, the risk is not minimal, nor is it a minor increment above minimal risk. The risk is substantial, and in initial trials, it is very likely that child participants will die soon after transplant. Hence, proceeding with this research in children before obtaining more data on safety and efficacy in adults first is controversial.

While pediatric xenotransplantation would not pass the test of minimal risk, we could also raise the question of whether it provides, at least initially, the potential for individual benefit. What exactly does *benefit* mean? The term on its own is ill-defined unless outcome markers are attached to it. For instance, does benefit mean that the cardiac xenograft allows the patient to live with the heart into adulthood? What about ten years, one year, or one month? Would those markers constitute a benefit? Would benefit mean that the cardiac xenograft provides the potential to "bridge" the patient until a future point when a human heart becomes available. After all, in the transplant context, it has been hypothesized that cardiac xenotransplantation could serve as a "bridge therapy" until a deceased human organ becomes available to serve as a "destination therapy."

However "benefit" is defined, it is certain that there is no guarantee of individual benefit for early phase clinical trials of pediatric cardiac xenotransplantation. What can be said is that there is *potential* for individual patient benefit. For instance, there is *potential* that cardiac xenotransplantation could provide the benefit of extending the life of the recipient. Yet, many unknowns also coexist with this potential for benefit. For example, it is unknown whether the potential for benefit is offset by the potential for harm, which could include pain for the child due to the operation (pain that the child is likely unable to understand), additional procedures that may ensure posttransplant, sensitization so that receiving a human heart transplant in the future becomes more difficult, and even death (Hurst et al., 2024). Hence, it is undecided how an appropriate risk-to-benefit assessment can be made.

There is an additional rationale that has been provided as to why children may especially be good candidates for cardiac xenotransplantation. Immunological advantages may exist for infants accepting a xenograft compared to adults (Bikhet et al., 2020). During the initial 6–12 months of life, infants possess minimal or absent preexisting anti-pig antibodies against organs derived from gene-edited pigs, which are commonly considered as potential sources for heart transplants. Furthermore, infants typically demonstrate less active complement systems and diminished innate immune cell responses. These factors are anticipated to lower the likelihood and intensity of rejection of xenografted hearts, especially among neonates who have limited mechanical circulatory support alternatives available.

So, we are left with a tension. On the one hand, while children are deserving of special protection in virtue of their vulnerability, they are likely to die waiting for a heart transplant: needing a pediatric heart is a death sentence for most. Yet, participating in xenotransplant research will likely also lead to death, just as it likely will lead to death for initial adult participants. On the other hand, children would seem to have a more favorable risk-benefit analysis compared to adults in virtue of their less active complement systems and diminished innate immune cell responses.

Perhaps the answer is that more preclinical research is required. Prior to proceeding to xenotransplantation clinical trials in adults, researchers have chosen to proceed with studies in both decedents (persons declared dead by neurologic criteria) as well as pursuing expanded access authorizations (i.e., compassionate use). In adults, both kidney (Porrett et al., 2022; Locke et al., 2023; Montgomery et al., 2022) and heart xenotransplantation (Moazami et al., 2023) have been studied in order to gather data on how the xenografts perform in a human body, albeit a deceased human body. Advantages are claimed to each of these pathways. It must be considered whether such approaches would be appropriate in pediatrics prior to beginning clinical trials. The main advantage claimed by researchers supportive of the decedent model is that they are able to gather valuable data that they are not able to gather from pig-to-baboon studies and do so in a manner that does not place living persons at risk (Montgomery et al., 2024). While there have not been survey studies of the public's view on xenotransplantation decedent studies, the absence of public backlash points in a direction of acceptance. However, we must remember that the persons used thus far for decedent studies have been adults—typically older adults, with one kidney xenograft decedent recipient being 57 years of age (Porrett et al., 2022), and cardiac xenograft decedents being 64 and 72 (Moazami et al., 2023). Hence, these are individuals who have lived a lot of life. However, the use of children who have been declared dead by neurologic criteria is different. First, approaching the grieving parent(s) of a child who has just died about whether they will consent to xenotransplantation raises apprehensions based on the intense emotional situation and distress that parents may be under. It may not be sensitive to approach parents with a request to use their child for xenotransplantation research. Second, it is uncertain how the public would feel about this line of research, and it is possible this could cause backlash.

Rather than decedent studies in pediatric xenotransplantation, expanded access authorizations may be the approach that researchers take. In the United States, the FDA requires that each of the following criteria must be met for expanded access to *potentially* be appropriate:

1. Patient has a serious or immediately life-threatening disease or condition.
2. There is no comparable or satisfactory alternative therapy to diagnose, monitor, or treat the disease or condition.
3. Patient enrollment in a clinical trial is not possible.
4. Potential patient benefit justifies the potential risks of treatment.
5. Providing the investigational medical product will not interfere with investigational trials that could support a medical product's development or marketing approval for the treatment indication. (U.S. Food and Drug Administration, 2024).

Taken together, these five criteria offer a compelling, ethically defensible position for proceeding with pediatric cardiac xenotransplantation expanded access studies possibly as a precursor to clinical trials. Criterion 1 can easily be met due to the situation that many pediatric cardiac patients are in. Simply put, many children with congenital heart defects require transplantation. Without a heart transplant, they will die. As mechanical circulatory support is limited in children, especially in infants, no comparable therapy to transplantation is available, which meets criterion 2. As of this writing (mid-2024), no xenotransplantation clinical trials—let alone pediatric cardiac xenotransplantation clinical trials—are underway (criterion 3). Even should a pediatric cardiac xenotransplantation clinical trial begin, this would not necessarily preclude offering pediatric cardiac xenotransplantation under expanded access guidelines because it may not be possible to enroll the patient in question into the clinical trial due to travel, financial considerations, caregiver support, etc. Criterion 4 can be met if researchers earnestly believe that the potential for benefit from the cardiac xenotransplantation justifies the risk. When the outcome is likely to be no transplant with high probability of mortality, meeting this threshold may not be difficult. Lastly, as no investigational trials are currently underway, criterion 5 is met.

Public viewpoints on pediatric xenotransplantation

The ethical issues attendant with pediatric xenotransplantation are distinct from, yet related to, the public's viewpoints about pediatric xenotransplantation. Hence, when discussing the ethics of pediatric xenotransplantation, it is important to examine the public's view. To date, there has been limited empirical data exploring the viewpoints of the public and, importantly, parents of children with congenital heart defects, toward xenotransplantation. Public viewpoint is important. Circling back to the Baby Fae xenotransplantation case in 1984, it was this event that brought many ethical issues of xenotransplantation to the foreground. Indeed, the case was criticized most notably by a leading Catholic moral theologian, Gino Concetti, for it offering little foreseeable benefit (Vatican Expert Views Surgery on Baby Fae, 1984; Rodger et al., 2023).

Since Baby Fae, a limited number of studies have explored viewpoints toward pediatric xenotransplantation. In a survey of pediatric congenital heart disease surgeons, pediatric transplant cardiologists, pediatric cardiac nurses, and parents of children with congenital heart disease, acceptance toward xenotransplantation was high overall (75%), yet this only held true if outcomes were likely to be similar to those of allotransplantation, which is unlikely to be the case in initial years of clinical xenotransplantation (Padilla et al., 2021). Parents generally had a less positive view of xenotransplantation than physicians or nurses (64%, 86%, and 71%, respectively). Furthermore, individuals who cited religion as a guiding factor in their medical decision-making tended to express lower levels of approval toward xenotransplantation. Another noteworthy aspect highlighted in this study is the expression of parental worries regarding the potential social impacts of xenotransplantation on their child's interactions with others. There appeared to be a correlation between higher levels of parental psychosocial concerns and reduced acceptance of the procedure. Similarly, in several focus group studies that included parents of children with congenital heart disease, participants voiced concerns about social stigma, expressing worries that their children could face bullying or teasing due to having a pig organ (Hurst et al., 2021).

In another survey of viewpoints toward pediatric xenotransplantation conducted among members of the Congenital Heart Surgeons Society and Pediatric Heart Transplant Society, comprising pediatric heart surgeons and cardiologists, the results revealed widespread acceptance of xenotransplantation provided that risks and outcomes were projected to be comparable to those of allotransplantation (Padilla et al., 2020). However, surprisingly, only 41% of surgeons and 17% of cardiologists would advise parents to consider a pig cardiac xenograft as a temporary measure leading to allotransplantation. Equally unexpected, both groups agreed that they would not recommend replacing the xenograft with a human heart, assuming a heart became available at a later point, if a pig cardiac xenotransplant resulted in favorable outcomes.

Hence, while there is limited empirical evidence on viewpoints of persons toward pediatric xenotransplantation, the data that exists raises concerns. Generally, there has not been much negative media coverage of the recent xenotransplantations that have taken place. Yet, xenotransplantation in children—a particularly vulnerable population—is likely to be put under greater ethical and medical scrutiny, especially if the outcomes are sub-optimal. Arguably, undertaking xenotransplantation in the pediatric population risks having a more deleterious effect on public perception. One must also consider the psychosocial consequences for the family and the long-term implications for the reputation of xenotransplantation research. The family of the child may be expected to give an account for their choice in a way that the family of an older patient would not be. This further justifies taking a cautious case-by-case approach because it must be assumed that the surgery will garner significant media and public attention and will be subject to ongoing debate—much like Baby Fae's case was. For xenotransplantation to be a successful therapeutic option, it is essential that the public does not become disillusioned and hostile toward it. Involving children will inevitably involve an increased risk of this, and the fallout from "Baby Fae" in 1984 serves as a warning.

Psychosocial considerations

It is largely speculative concerning how xenotransplantation may affect a person psychosocially. However, there are some indications from the allotransplantation community that are worth considering. From the allotransplantation literature, it is known that many psychosocial factors have the ability of impacting how a person adjusts posttransplant and may impact clinical outcomes (Randolph et al., 2023). Psychological and social assessment is typically a standard part of pretransplantation assessment with some standardized tools available for conducting this evaluation, though this is an area that is still largely subjective (Bui et al., 2019; Dew et al., 2018; Fidel Kinori et al., 2015). For obvious reasons, psychological and social assessments are more complex for children.

With any type of transplantation—from a human or genetically modified pig—nonadherence to the prescribed immunosuppression regimen is a major risk factor for poor outcomes. Immunosuppression medications are needed in allotransplantation and will also be required for xenotransplantation, as they lower the body's natural ability to reject a foreign tissue, such as an organ. In pediatric allotransplantation, it has been well-documented that nonadherence to the prescribed immunosuppressive regimen is a common phenomenon (Dobbels et al., 2005). The prevalence of nonadherence among pediatric allotransplant recipients has been calculated at 32% for kidney transplantation, 31% for liver, and 16% for heart. Researchers have postulated several risk factors or determinants of nonadherence to immunosuppressive regimen, including demographic and socio-economic factors like family instability or lack of social support, patient-related factors such as poor understanding of their disease and the importance of lifelong immunosuppression, and treatment-related factors like the cosmetic side effects that immunosuppression medications may cause, a factor that may be more significant to adolescents (Dobbels et al., 2005).

Researchers have recommended that in order for xenotransplantation to have a chance of success, selecting candidates will require careful psychosocial consideration. This is to ensure that the patient has adequate decision-making capacity and screening for mental health disorders (Randolph et al., 2023). While this may be possible in adults and older children, this is not possible for infants or younger children. Hence, it may be necessary to follow these children for several years post-xenotransplantation to ensure they have the psychosocial support needed.

Informed consent and lifelong monitoring

One area we have discussed elsewhere in this book that needs attention in the pediatric setting is the issue of lifelong biosurveillance monitoring. To review, several commentators have recommended that xenograft recipients agree to lifelong monitoring for any xenozoonotic disease. While an adult may be able to consent to this, a child—especially a very young child—cannot. Hence, if lifelong biosurveillance is a requirement, then a parent/guardian must provide surrogate consent on behalf of the child. This would entail that, even into adulthood, the recipient may not be able to withdraw from this specific portion of the clinical trial, which they never agreed to in the first place, for a very long time (Hurst et al., 2020). This may limit the child's future options, such as the ability to live wherever they wish if they must live near a medical center that can perform the biosurveillance.

Informed consent in the context of unknown xenozoonotic risks presents complex challenges, including (i) how to approach consent or assent from the recipient and (ii) identifying third parties who are obligated to consent on behalf of the patient (Hurst et al., 2024). Further deliberation is needed on these aspects, including how best to counsel surrogate decision-makers who are consenting on behalf of the recipient. Given that parents or guardians may be in a desperate situation and willing to try any available means to save their child's life, careful attention must be given to ensure they fully understand the *research* nature of xenotransplantation, the risks involved (including unknown risks), and the potential need for subsequent heart re-transplantation, whether the xenograft is used as a bridge or fails.

Parents and guardians should also be informed of the potential impacts of participation in initial trials on their privacy and that of their child, including possible implications for health insurance. The consent process must be free from industry and researcher influence. Since parents or guardians may place undue trust in the treating physician's recommendations, it may be beneficial to have consent obtained by a qualified individual who is not directly involved in the child's care.

Other considerations

Other considerations of the "known unknowns" variety that ought to be discussed are the burdens of potential stigmatization from their peers or community members that may come earlier or later in life; the potential attitudes toward animal welfare the child may come to have over time; and, assuming the xenograft works well over the long term, what implications there might be for having children themselves. Of course, the relevance these issues have for decision-making will depend on what stage xenotransplant research has progressed to, and whether there are legitimate clinical options to select. Nonetheless, the risk of stigmatization, which was mentioned briefly above, may be the most relevant as it would seem to be present at any time in one's life. By contrast, the risks associated with imposed value judgments about animals or the limits on childbearing may be less relevant at the time of this writing since the potential for experimental xenografts to increase both the quantity and quality of life may be rather modest.

Conclusion

Pediatric xenotransplantation has come a long way since Baby Fae in 1984. However, the field is only now getting to the point of potentially beginning genetically modified pig-to-human clinical trials. Though there is a great need for healthy, transplantable organs (namely hearts) in the pediatric population, argument persists regarding whether xenotransplantation may proceed in children before first ensuring safety and efficacy in adults. While we understand the rationale, the totality of the situation must be accounted for with the brunt truth that children will continue to die due to lack of alternative cardiac support, which xenotransplantation may be able to provide. There is an urgent need for clarity on many of these issues.

Let us review our progression through the ethics of xenotransplantation. After discussing the need for organs for transplant (Chapter 1), we engaged the animal and public health ethics of genetically engineering pigs for their organs (Chapters 2 and 3). Assuming a favorable analysis of these issues, we discussed the ethics of designing xenotransplant clinical trials and future recipient selection criteria (Chapter 4) and the difficulties surrounding securing informed consent in adult (Chapter 5) and pediatric (this chapter) populations. It is important to keep in mind that at every step along the march to xenotransplant clinical reality are the biotechnology companies that employ and fund xenotransplant research. The topic of big business in xenotransplantation is the focus of our next chapter.

References

Bailey, L. L. 2011. Origins of neonatal heart transplantation: An historical perspective. *Seminars in Thoracic and Cardiovascular Surgery: Pediatric Cardiac Surgery Annual*, 14, 98–100.

Bikhet, M., Morsi, M., Hara, H., Rhodes, L. A., Carlo, W. F., Cleveland, D., Cooper, D. K. C. and Wase, H. 2020. The immune system in infants: Relevance to xenotransplantation. *Pediatric Transplantation*, 24, e13795.

Blume, E. D., Vanderpluym, C., Lorts, A., Baldwin, J. T., Rossano, J. W., Morales, D. L. S., Cantor, R. S., Miller, M. A., St Louis, J. D., Koehl, D., Sutcliffe, D. L., Eghtesady, P., Kirklin, J. K. and Rosenthal, D. N. 2018. Second annual pediatric interagency registry for mechanical circulatory support (Pedimacs) report: Pre-implant characteristics and outcomes. *Journal of Heart Lung Transplant*, 37, 38–45.

Bui, Q. M., Allen, L. A., Lemond, L., Brambatti, M. and Adler, E. 2019. Psychosocial evaluation of candidates for heart transplant and ventricular assist devices: Beyond the current consensus. *Circulation: Heart Failure*, 12, e006058.

Council for International Organizations of Medical Sciences. 2016. *International Ethical Guidelines for Health-Related Research Involving Humans*. https://cioms.ch/wp-content/uploads/2017/01/WEB-CIOMS-EthicalGuidelines.pdf (Accessed: 16 January 2024).

Dew, M. A., DiMartini, A. F., Dobbels, F., Grady, K. L., Jowsey-Gregoire, S. G., Kaan, A., Kendall, K., Young, Q. R., Abbey, S. E., Butt, Z., Crone, C. C., De Geest, S., Doligalski, C. T., Kugler, C., McDonald, L., Ohler, L., Painter, L., Petty, M. G., Robson, D., Schlöglhofer, T., Schneekloth, T. D., Singer, J. P., Smith, P. J., Spaderna, H., Teuteberg, J. J., Yusen, R. D. and Zimbrean, P. C. 2018. The 2018 ISHLT/APM/AST/ICCAC/STSW recommendations for the psychosocial evaluation of adult cardiothoracic transplant candidates and candidates for long-term mechanical circulatory support. *The Journal of Heart and Lung Transplantation*, 37, 803–823.

Dobbels, F., Damme-Lombaert, R. V., Vanhaecke, J. and Geest, S. D. 2005. Growing pains: Non-adherence with the immunosuppressive regimen in adolescent transplant recipients. *Pediatric Transplantation*, 9, 381–390.

Dommel, F. W. 1986. The NIH report of its review of the baby Fae case. *IRB: Ethics & Human Research*, 8, 1–4.

Fidel Kinori, S. G., Alcántara Tadeo, A., Castan Campanera, E., Costa Requena, G., Diez Quevedo, C., Lligoña Garreta, A., Lopez Lazcano, A., Martínez del Pozo, M., Rangil Muñoz, T. and Peri Nogués, J. M. 2015. Unified protocol for psychiatric and psychological assessment of candidates for transplantation of organs and tissues, PSI-CAT. *Revista de Psiquiatría y Salud Mental*, 8, 130–136.

Hamzelou, J. 2023. This company plans to transplant gene-edited pig hearts into babies next year. *MIT Technology Review*. https://www.technologyreview.com/2023/07/17/1076392/this-company-plans-to-transplant-pig-hearts-into-babies-next-year/ (Accessed: 5 March 2024).

Health Resources and Services Administration. *Organ Donation and Children*. https://www.organdonor.gov/learn/who-can-donate/children (Accessed: 30 August 2024).

Hurst, D. J. and Padilla, L. 2024. Ethically advancing pediatric cardiac xenotransplant. *JAMA Pediatrics*, 178, 5–6.

Hurst, D. J., Padilla, L. A., Cooper, D. K. C. & Paris, W. 2021. Factors influencing attitudes toward xenotransplantation clinical trials: A report of focus group studies. *Xenotransplantation*, e12684.

Hurst, D. J., Padilla, L., Merlocco, A., Rodger, D., Bobier, C., Gray, W. H., Sorabella, R., Cooper, D. K. C. and Pierson, R. N. 2024. Pediatric cardiac xenotransplantation: Recommendations for the ethical design of clinical trials. *Transplantation*, 108(10), e292–e300.

Hurst, D. J., Padilla, L. A., Walters, W., Hunter, J. M., Cooper, D. K. C., Eckhoff, D. M., Cleveland, D. and Paris, W. 2020. Paediatric xenotransplantation clinical trials and the right to withdraw. *Journal of Medical Ethics*, 46, 311–315.

Jonasson, O. and Hardy, M. A. 1985. The case of baby Fae. *JAMA*, 254(23), 3358–3359.

Knoll, E. and Lundberg, G. D. 1985. Informed consent and baby Fae. *JAMA*, 254(23), 3359–3360.

Locke, J. E., Kumar, V., Anderson, D. and Porrett, P. M. 2023. Normal graft function after pig-to-human kidney xenotransplant. *JAMA Surgery*, 158, 1106–1108.

Moazami, N., Stern, J. M., Khalil, K., Kim, J. I., Narula, N., Mangiola, M., Weldon, E. P., Kagermazova, L., James, L., Lawson, N., Piper, G. L., Sommer, P. M., Reyentovich, A., Bamira, D., Saraon, T., Kadosh, B. S., DiVita, M., Goldberg, R. I., Hussain, S. T., Chan, J., Ngai, J., Jan, T., Ali, N. M., Tatapudi, V. S., Segev, D. L., Bisen, S., Jaffe, I. S., Piegari, B., Kowalski, H., Kokkinaki, M., Monahan, J., Sorrells, L., Burdorf, L., Boeke, J. D., Pass, H., Goparaju, C., Keating, B., Ayares, D., Lorber, M., Griesemer, A., Mehta, S. A., Smith, D. E. and Montgomery, R. A. 2023. Pig-to-human heart xenotransplantation in two recently deceased human recipients. *Nature Medicine*, 29, 1989–1997.

Montgomery, R. A., Griesemer, A. D., Segev, D. L. and Sommer, P. 2024. The decedent model: A new paradigm for de-risking high stakes clinical trials like xenotransplantation. *American Journal of Transplantation*, 24, 526–532.

Montgomery, R. A., Stern, J. M., Lonze, B. E., Tatapudi, V. S., Mangiola, M., Wu, M., Weldon, E., Lawson, N., Deterville, C., Dieter, R. A., Sullivan, B., Boulton, G., Parent, B., Piper, G., Sommer, P., Cawthon, S., Duggan, E., Ayares, D., Dandro, A., Fazio-Kroll, A., Kokkinaki, M., Burdorf, L., Lorber, M., Boeke, J. D., Pass, H., Keating, B., Griesemer, A., Ali, N. M., Mehta, S. A. and Stewart, Z. A. 2022. Results of two cases of pig-to-human kidney xenotransplantation. *New England Journal of Medicine*, 386, 1889–1898.

Nairne, P. and Nuffield Council on Bioethics (eds.). 1996. *Animal-to-Human Transplants: The Ethics of Xenotransplantation*. London: Nuffield Council on Bioethics.

Padilla, L. A., Rhodes, L., Sorabella, R. A., Hurst, D. J., Cleveland, D. C., Dabal, R. J., Cooper, D. K., Paris, W. and Carlo, W. F. 2021. Attitudes toward xenotransplantation: A survey of parents and pediatric cardiac providers. *Pediatric Transplantation*, 25, e13851.

Padilla, L. A., Sorabella, R., Carlo, W., Dabal, R., Rhodes, L., Cleveland, D., Cooper, D. and Paris, W. 2020. Attitudes to cardiac xenotransplantation by pediatric heart surgeons and physicians. *World Journal for Pediatric and Congenital Heart Surgery*, 11, 426–430.

Porrett, P. M., Orandi, B. J., Kumar, V., Houp, J., Anderson, D., Cozette Killian, A., Hauptfeld-Dolejsek, V., Martin, D. E., Macedon, S., Budd, N., Stegner, K. L., Dandro, A., Kokkinaki, M., Kuravi, K. V., Reed, R. D., Fatima, H., Killian, J. T., Jr., Baker, G., Perry, J., Wright, E. D., Cheung, M. D., Erman, E. N., Kraebber, K., Gamblin, T., Guy, L., George, J. F., Ayares, D. and Locke, J. E. 2022. First clinical-grade porcine kidney xenotransplant using a human decedent model. *American Journal of Transplantation*, 22, 1037–1053.

Randolph, B., Nosker, J. and Jimenez, T. 2023. Psychological implications. In Hurst, D. J., Padilla, L. and Paris, W. D. (eds.), *Xenotransplantation: Ethical, Regulatory, and Social Aspects*. Cham: Springer International Publishing.

Rodger, D., Hurst, D. J. and Cooper, D. K. 2023. Xenotransplantation: A historical-ethical account of viewpoints. *Xenotransplantation*, 30, e12797.

U.S. Department of Health and Human Services. 1979. *The Belmont Report*. https://www.hhs.gov/ohrp/regulations-and-policy/belmont-report/index.html (Accessed: 5 March 2024).

U.S. Department of Health and Human Services. 2021. *Organ Procurement and Transplantation Network (OPTN) and Scientific Registry of Transplant Recipients (SRTR)*. OPTN/SRTR 2021 Annual Data Report. http://srtr.transplant.hrsa.gov/annual_reports/Default.aspx (Accessed: 5 March 2024).

U.S. Food and Drug Administration. 2024. *Expanded Access*. https://www.fda.gov/news-events/public-health-focus/expanded-access (Accessed: 16 May 2025).

Vatican expert views surgery on baby Fae. *The New York Times*, 1984, November 18.

7 How do we balance innovation and price?

The bad business CEO is a common villain trope in books, movies, and television shows. Usually, but not always, the villain is trying to make a handsome profit at the expense of vulnerable people, and given high prices in healthcare and the promise of biotechnology, readers and viewers alike are not surprised when the villain is the CEO of a biotechnology company. To offer but one example, according to the storyline of the 2010 movie, *Repo Men*, starring actors Jude Law and Forest Whitaker, a biotechnology company sells biomechanical organs to people in the year 2025. Since the organs are expensive, recipients agree to pay for the organ over time, that is, in monthly payments. Of course, life happens and recipients sometimes stop payment, for either nefarious or benign reasons—it is mostly the latter, usually due to unexpected financial hardship. When recipients stop payment, the biotechnology company sends in ruthless repossession men to retake the organs, resulting in the death of the recipient. Amidst the action and character development, the movie presents a simplistic but widely shared understanding of the relationship between medicine and business: a biotechnology company's primary, if not sole, aim is to profit from technology that people need to survive.

Repo Men is correct that biotechnology companies are at the forefront of developing transplantable organs, albeit genetically engineered porcine organs, but the relationship between medicine and business is more ethically complex than what is portrayed in the movie. The research required to bring xenotransplantation technology to the bedside is expensive and requires significant funding, which is often not sufficiently covered by public funds. Once it becomes clinically implemented, xenotransplantation may be quite lucrative for biotechnology companies that have invested significantly in funding the research and development. It stands to reason, then, that companies will patent their inventions and charge a high premium. But patenting potentially lifesaving living organs raises a host of ethical questions. Do the benefits of awarding patents to biotechnology companies

DOI: 10.4324/9781003472223-8

outweigh the risks? Does patenting an organism with some human genetic material blur morally significant boundaries? Does society owe something, such as patents, to inventors and investors who invent something of social value? Also, patenting in xenotransplantation may severely restrict accessibility to potentially lifesaving organs if these companies charge an excessive fee, which they have reason to given the significant investment and goal of profiting. Does distributive justice suggest greater governmental control over the patented invention?

The aim of this chapter is to engage these and related ethical questions, questions that historically align with the field of business ethics. This further highlights our observation made in the Introduction that xenotransplantation raises questions that span ethics. As with previous chapters, but to be as clear as possible, we limit our discussion to the US context. We begin with a discussion of patents, their rationale, and their scope. Given legal patent criteria, we explain how gene-editing technology, genetically modified organs, and entire pig lines have been patented. Then, we engage the ethics of patenting in xenotransplantation, highlighting the case in its favor and criticisms. Finally, we discuss whether the US government should intervene in patent protections to force biotechnology companies to lower costs.

Xenotransplant patents

According to the United States Constitution, the purpose of a patent is to "promote the progress of science and useful arts, by securing for limited times, for authors and inventors, the exclusive rights to their respective writings and discoveries" (US Const. Art1.s8.cl8). Upon conferral, a patent grants legal monopoly by the government to the inventor, which allows the inventor to prohibit others for a set time (20 years) from making, using, or selling the invention. A patent does not convey any positive right to make, use, or sell an invention; it confers only a negative right of exclusion: the patent holder can prevent others from making, using, or selling the invention. It is possible for a patent holder to do nothing with the invention, to not profit from it or even produce it.[1] The reason for granting patents, that is, special privilege to a proprietary invention, is to spur progress that is beneficial to society. Congress' aim is to foster "the productive effort" and thereby benefit "society through the introduction of new products and processes of manufacture" (US Const. Art 1.s8, cl8). Why invest in inventing something new if someone else can take it and profit from it, leaving you in the dust? Without patent protection, smaller companies and research organizations may have their inventions capitalized on by larger organizations. But there is more to the rationale for patents: patents require public disclosure of information regarding the design of the invention, thereby allowing for greater transparency, accountability, and rigor.

The courts have conditions on what can and cannot be patented. In general, things found in nature, even things that require sustained investigation to discover, are not patentable. If a researcher discovers a hitherto hidden planet or a law of nature, she cannot patent the planet or the law of nature. Patentable subject matter includes "anything under the sun that is *made by man*" (*Diamond v. Chakrabarty*, 447 US 303, 309). A rough litmus test would be whether the object or organism is a mere discovery (i.e., product of nature) or a man-made invention (i.e., product derived from nature). It is important to keep in mind that objects or organisms (i.e., products) are distinct from the process that results in their existence (i.e., process): the former refer to "tangible items," while the latter refer to the steps that have to be undertaken to produce a specific object (*In Re John Kollar, 286 F.3d 1326*, 1332). Consider Dolly the Sheep, the first cloned sheep in 2009. While the United States courts granted a patent on the *process* of cloning used to create Dolly, the courts denied that Dolly and other products of that process were patentable on grounds that Dolly was a product of nature: "Dolly herself is an exact genetic replica of another sheep and does not possess 'markedly different characteristics from any [farm animals] found in nature'" (*In Re Roslin Inst. (Edinburgh)* 750 F 3d 1333, 1337). While sheep exist in nature, which is why Dolly, a genetically cloned sheep, could not be patented, the process of making Dolly does not naturally occur in nature, and so, the process of cloning was patentable but not the product of cloning.

United States courts have set four legal conditions that would-be patents must meet (Lim and Li, 2022; Lever, 2008). First, the subject matter is limited to a "process, machine, manufacture, or composition of matter" (US Code, Title 35, §101). This condition rules out patents on immaterial things such as ideas or discoveries of nature, such as new planets in the solar system. Second, patented materials have to be novel or new, that is, not existing in nature. A novel process of creating genetically identical-to-recipient organs would be patentable, but not the product of the process, just as Dolly the Sheep was not patentable but the process of creating her was. Similarly, the invention cannot be contained in a prior patent—it has to be a new invention. Third, the invention has to be "nonobvious," meaning it cannot be a mere extension of already existing technology: slight modifications to CRISPR gene-editing technology cannot be patented, but a new gene-editing technology or significant, nonobvious improvement upon CRISPR can be patented. "Nonobvious" is defined by those familiar with or trained in the relevant invention area. Finally, the invention has to be useful or beneficial to humankind. United States Code Title 42, § 2181, for instance, revokes and denies future patents for "any invention or discovery which is useful solely in the utilization of special nuclear material or atomic energy in an atomic weapon." Still, many weapons are patented (e.g., safer and simpler cluster bomb–patent # US 4744301A), so there is ambiguity (to put it mildly) about what "useful" or "beneficial" means.

United States courts have granted patents to a variety of organisms or parts of organisms (e.g., altered DNA sequences) in addition to cloning techniques and *in vitro* fertilization techniques. In *Diamond v. Chakrabarty 447,* the United States Supreme Court ruled that an oil-eating bacterium was the result of "human ingenuity," for it was a "non-naturally occurring manufacturer or composition of matter" (309). Following the *Chakrabarty* case, a patent was granted for the "Harvard Oncomouse," which, as Lim and Le explain, is "a transgenic non-human mammal with an activated oncogene sequence introduced at embryonic stage to predispose the mice to cancer that was suitable for carcinogenic studies" (2022, p. 72). The invention was new, useful, nonobvious, and, given the complexity involved, not naturally occurring: it is not as though one discovers such mice in the wild. In addition, particular modified human gene sequences have been patented. For instance, the courts have held that synthetic complementary DNA, or DNA that has been modified with some proteins removed, can be patented, as modification through extraction and addition results in something that does not already exist in nature (Cartwright-Smith, 2014).

Human beings are obviously not patentable, as we are naturally occurring organisms. While this position had been implicit for a long time, it was stated clearly in the 1990s and early 2000s. In a policy statement, the United States Patent and Trademark Office (PTO) stated that people could get patents for non-naturally occurring, nonhuman organisms (Goodman, 2017), and then in 2004, the United States Congress enacted a measure that prohibits patent claims that are "directed to or encompassing a human organ" (The Consolidated Appropriations Acts of 2004. Pub L No 108–109, 118 Stat. 3). While there have been patents granted on genetically modified animals that have some human genes, not to mention patents on human cells and tissues, the PTO rejected a patent application for a human-animal chimera that could be up to 50% human. In the 1990s, two biotechnology activists applied for a patent on a variety of human-animal combinations—their goal was to force further discussion on patenting living creatures, including creatures that are part-human. O'Sullivan explains that the "PTO rejected the application, opining that the inventions resemble human beings too closely to be patentable" (2018, p. 66). It was left vague what "resemble" means: phenotypic resemblance, genetic resemblance, or some combination?

Given the proliferation of patents in animal agriculture and research, there has been and continues to be no legal opposition to patenting in xenotransplantation. Clearly, xenotransplant research satisfies the four legal conditions. The subject matter is limited to a process or composition of matter, a genetically engineered porcine family, organ, or process of genetic alteration. The genetically engineered organs and host animals are not existing in nature, and they require nonobvious, highly technical engineering to produce. The inventions promise to be useful to society: a new source of transplantable organs and other lines of research. Because xenotransplantation research involves the insertion of select human genes into the host species, xenotransplant patents do not pose a problem with the prohibitions against patenting human beings. In other words, the genetically engineered pig with human genes does not resemble human beings enough to cause worry, although some may object to this very point.

Still, there is the issue of what is patentable. There are three possibilities: the process of creating the genetically engineered pig, the transplantable organ, or the genetically engineered pig. The PTO has shown willingness to patent each in xenotransplantation research, as evidenced by the following (Holzer et al., 2024):

- Process: A patent granted to Mayo Foundation for Medical Education and Research regarding the methods and materials for creating a certain line of transgenic pigs.
- Organ: A patent granted RBC Biotechnology for the rights to a line of modified swine organs created to be more suitable for human transplantation.
- Pig and pig line: A patent granted to The Trustees of Columbia University (New York) regarding the donor animals that carry a specific set of genetic alterations.

Businesses and researchers have a vested interest in patenting the genetically engineered animal line, for this would include the organ and process of creating the organ as well. It is not uncommon to see patent applications include the method of creating the transplantable organ, the host animal species, and the organ. For instance, biotechnology company Revivicor received a patent on genetically engineered animals that have certain genetic alterations that render them more suitable for transplantation: the patent includes the animals, their organs, and the "methods of making such animals" (Holzer et al., 2024, p. 30).

Ethics of xenotransplant patents

While the ethics of xenotransplant patents have yet to be discussed in detail in the scholarly literature, there are a number of ethical questions that can be raised. Do the good consequences of patenting xenotransplants outweigh the bad consequences? Are inventors owed something, for example, patents, for their socially beneficial inventions? Is it the first step on a slippery slope to undermining human dignity? Many of these questions are not unique to xenotransplantation, but they are integral to a thoughtful consideration of the relation between business and xenotransplantation. Our goal in this section is to present the ethical pros and cons of the issue. We divide our discussion into two sections, one that focuses on arguments attending to the consequences of xeno-patenting and one that focuses on non-consequentialist arguments.

Consequentialist arguments

When examining the ethics of a policy, a helpful starting place is to consider the policy's risks and benefits, and to contrast these with alternative(s) (Lever, 2008; United States Congress, 1987). When doing so, it would seem that proponents of xeno-patents have a strong *prima facie* case for allowing patents: the practice promises to contribute to highly desirable consequences with no offsetting bad consequences. Patenting the production of xenotransplants creates a financial incentive for their creation, which is important given that medical research is costly and time-consuming. In pharmacology, for instance, it takes around ten years for a medicine to receive regulatory approval and may cost more than $1 billion to research and develop (Wouters et al., 2020). In exchange for such significant financial and time investment, the biotechnology company receives legal protection on the patented process or product invented, thereby promising a potentially significant financial return on investment. Given that Congress' aim is to foster "the productive effort" and thereby benefit "society through the introduction of new products and processes of manufacture" (US Const. Art 1.s8, cl8), xeno-patents are valuable. Without such protections, it might be wondered if biotechnology companies would invest as much time and money into the research and development of xenotransplant organs.

To support the consequentialist argument in favor of xeno-patents, three considerations should be assessed (Resnik, 2012). *First, how desirable is the invention?* For reasons discussed elsewhere in the book, there is a broad consensus that a new source of transplantable organs is desirable: many more people are in need of a solid organ transplant than there are organs available for transplantation. There are concerns about equitable distribution of xeno-organs, both within the United States and internationally, not to mention concerns of zoonotic infection, and critics may insist that harvesting genetically engineered porcine organs is the wrong solution to the problem. Nevertheless, defenders of xeno-patents think the potential good of a new source of transplantable organs outweighs the potential cons. *Second, how likely is the invention to be achieved?* For reasons discussed elsewhere in the book, there is reason for optimism that clinical trials will begin for kidney, liver, and heart xenotransplantation within the next decade or so. *Third, how likely is the desired outcome to be achieved in the absence of patent protection?* This is difficult to assess since xeno-patents have been granted thus far with little resistance. Despite a general dearth of evidence (Williams, 2017), Sellars argues that there is "ample evidence to support the proposition that transgenic research will subside as a result of lost economic incentives if animal patent barriers are erected" (1994, p. 283) and there is evidence that patent protections spur research at smaller organizations as well as follow-on research (Arora, 2023; Kanwar and Evenson, 2003).

There are two further consequentialist considerations in favor of xeno-patent protections in the United States. First, patent protection helps keep the United States competitive in the international biotechnology arms race. By incentivizing businesses and researchers, not to mention permitting the research to continue, xeno-patents help ensure the United States is at the forefront of what may promise to be a significant new technology for the world. This is important for a number of United States-facing reasons (The White House Office of Science and Technology Policy, 2023): it allows the government to regulate biotechnology to protect public health; it helps keep a large, lucrative industry in the country; relatedly, it helps prevent "brain drain," or the export of researchers to other countries; the government has a vested interest in technology that promotes health and prevents death; and finally, xeno-patents may spur further downstream research in genetic engineering, research that may have untold positive benefit. Second, patents contribute to the public availability of knowledge. Patents allow inventors to publicly explain and claim their invention without fear that the invention will be stolen. Without patent protection, companies and researchers may resort to keeping their technology hidden, as a set of trade secrets. This in turn may hinder research and safety, as the technology is kept secret from researchers who can critically assess it. Patent protections promote research transparency, which in turn promotes safer and more responsible research.

The consequentialist case for xeno-patents seems pretty compelling, right? Well, the issue is not as clear-cut as defenders of the practice assume. Consider zoonotic risk, a topic that has been discussed in many places in this book already. Patenting this line of research, the process and product, promotes this line of research, which may in turn increase the risk of public harm resulting from potential zoonotic disease. To refuse patent protections would disincentivize this line of research, which in turn may decrease or slow xenotransplant research and thereby lower potential public health risk. If it is responded that the research would *not* cease if patents were no longer issued and existing ones revoked, then a primary rationale for xeno-patents loses force: if the research would continue in the absence of xeno-patents, then why posit the need of xeno-patents? Xeno-patents do not spur innovation in xenotransplant research after all. Since the topic of zoonosis has been discussed at length elsewhere in the book, we set it aside here but encourage readers to continue to think about how or whether the potential zoonotic risk affects the risk-benefit analysis of xeno-patents.

There are other issues to keep in mind when considering the consequences of xeno-patenting. One concern is that xeno-patents compromise research by blurring academic or social interests with business interests. Researchers who work at biotechnology companies pursue research, not solely for the improvement of the human condition, but to make a profit. This is not to say that they do not care about improving human life—they clearly do. Rather, our point is that their career depends, in large part, on their ability to research and develop something that can be monetized: no money, no job. This, in turn, incentivizes research on profitable technologies, which is not always the same as technologies that promise to help humanity. If a new technology promises to save countless lives but is unlikely to be profitable, biotechnology companies do not have an interest to invest significant resources and time into developing it.

This consideration is not limited to only researchers who work at biotechnology companies. It is well established that corporate funding influences the direction of research at universities (Fabbri et al., 2018). This makes sense. If a sugar company offers a researcher a sizable grant to research the correlation between sitting long periods and risk of heart disease, the researcher has financial reason to do so; this is so even though the sugar company will use the research to distract people from the well-established correlation between sugar intake and heart disease. In xenotransplant research, there is a close relationship between university researchers and biotechnology companies. The biotechnology company United Therapeutics Corporation awarded millions in funding to researchers—two Johns Hopkins Medicine surgeons received a total of $21.4 million in funding over two years under two sponsored research agreements with the company (Newman, 2023) and the University of Alabama at Birmingham (UAB) Comprehensive Transplant Institute was awarded a five-year, $19.5 million grant to launch a pioneering UAB Xenotransplantation Program (Greer, 2016). These researchers and universities now have a vested interest in researching xenotransplantation at the expense of other possible solutions to the organ supply deficit.

There is more. The introduction of a for-profit business model, spurred on by patent protections, incentivizes transplantation, not preventative measures that may lower transplant demand. The current donation-based system incentivizes public health measures, for there are not enough transplantable organs and society needs to lower the demand for them. (That said, there are powerful forces at play lobbying for and against measures that promote public health.) By introducing business incentives into the transplant system, there will be reasons to maintain the status quo and to resist measures that threaten profitability, even ones that promise to promote public health.

Another concern is that xeno-patents may hinder rather than spur research, and this is so for two reasons. First, xeno-patents may stifle collaboration and delay research findings. If researchers and organizations are working on a yet-to-be patented lucrative gene-edited porcine organ, they have every reason to keep this work from the public forum until after it has been patented. Since it takes, on average, over a year and a half for a patent to be awarded by the PTO, researchers may be worried that publishing research involved in a patent proposal may be swooped on by competitors. Better for researchers, but not necessarily society, to not publish until after they receive patent protections. Second, a proliferation of xeno-patents may threaten to slow down research, resulting in "multiple owners blocking each other through the proliferation of fragmented and overlapping intellectual property rights" (Joly, 2007, p. 389). This is referred to as a "patent thicket," which involves the maze of potential patent infringements that a researcher must navigate in order to develop a new invention. To disincentive competition, companies try to get as many patents as possible in the vicinity of a given technological innovation to slow down competitors. If multiple companies hold patents on xeno-related gene-editing processes, for instance, a researcher would need to obtain multiple different licenses to conduct her research. Larger biotechnology companies can more easily absorb licensing costs, but smaller companies and university researchers may not be able to easily navigate the legal maze. High licensing fees can slow down research and may result in prohibitively expensive-to-most invention.

A final set of considerations focuses on the impact xeno-patents may have on animals, including humans (Resnik, 2012; Lever, 2008). First, it may be argued that patenting living organisms, including pigs, opens the door to patenting human beings. This "slippery slope" argument goes as follows: patenting mammalian species with a small amount of human genetic material for the sake of organ harvesting opens the door to patenting mammalian species with slightly more human genetic material, and so on. Xenotransplantation of solid organs such as hearts, kidneys, and livers is desirable given the need for such organs; but gene-editing porcine brains to make them more human-like for the purposes of researching Alzheimer's is also highly desirable, but more worrisome. A humanized porcine brain created for the sake of research is, many think, more ethically worrisome given the close relationship between our brains, consciousness, and identity. Second, xeno-patents may increase animal suffering. Since the animal itself is patented, and patents only apply to nonnatural organisms, it is questionable whether genetically engineered pigs are legal animals. Stated differently, patents apply to inventions, and so, it is unclear whether patented organisms have protections under the Animal Welfare Act. "New entities covered by patent law," O'Sullivan writes, "are not treated very differently from their lifeless precursors" (2018, p. 64).

Non-consequentialist arguments

The previous set of arguments is all focused on the consequences of the xeno-patent practice, with plenty of empirical speculation being made. While not a criticism *per se* of the arguments, it does make one wonder whether non-consequentialist arguments can be offered, arguments that take a more principled stand. In this subsection, we present and evaluate one principled argument for each side.

Defenders of xeno-patenting have a principled argument, one that has to do with what is owed to researchers and companies that invent something of social benefit: "Inventors," Baruch Brody writes, "are entitled to patents as an acknowledgment of their efforts" (1989, p. 143). The idea is that society owes researchers and inventors for their socially valuable contributions (Lever, 2008). To motivate the idea, imagine someone who spends years trying to invent a new kind of battery, one that self-charges, has a significant charge, and requires much less natural resources to produce than current on-market batteries. This inventor owns the invention: she made it—it is the fruit of her labor. It would be wrong of someone else to take the newly invented battery and sell it as their own, for that would be stealing. Since people have a right to the fruit of their labor, it can be argued that researchers are entitled to the fruit of their intellectual labor. Just as you own the house you build from scratch on the land you own, so you own the invention you create. Since inventions may benefit others, society needs to find a way for inventors to share their inventions while retaining ownership. Patents are a compelling way of protecting intellectual property rights while balancing the social good of making inventions public—they give inventors their due while ensuring that others may benefit from the invention.

This argument can be scrutinized by noting that many of these organizations and researchers who receive patents could not have achieved this feat of research without the federal funding they received. David Cooper, a leading xenotransplant researcher, crunched the numbers and found that the US government, and the National Institutes of Health in particular, contributed a significant amount of financial support to xenotransplant research between 2000 and 2018, $60 million to be precise (2019, p. e12483). This falls short of the amount of money corporations spent on the research, which Cooper approximates to be around $100 million during this time period, but it does complicate the aforementioned line of thinking. If the government gives a researcher $1 million to invent a new battery, then that person's invention is not entirely the product of her own effort. Yes, the researcher put in the effort into producing it, but the research could not have been possible without the government funds. Does not the government have *some* claim on the invention?

It may be responded, however, that there is a distinction between enabling an achievement and accomplishing it. A parent who helps a child win a game plays a different role from the child playing the game. While the parent's role is important, the achievement of winning the game is largely due to the child's efforts (and plenty of luck). Similarly, funding for research enables research, but the results of research are primarily a function of the researcher (and plenty of luck). Moreover, the purpose of government funding is to spur socially beneficial research. When the government gives the $1 million grant to the researcher to create a new battery, it does not seek to own or otherwise claim the invention; instead, the government's interest is in having a new battery to market. Lastly, there is the concern that, if the government claimed co-ownership, say, over everything invented with the help of federal money, then researchers and biotechnology companies would not want to receive federal money. This may slow down inventions, thereby slowing down potential public benefit. These and related lines of thinking were codified into law in the passage of the Patent and Trademark Act Amendments of 1980, or Bayh-Dole Act, which allows inventors who invent something with the help of federal support to retain the patent rights (Eberle, 1999). (More on the Bayh-Dole Act later.)

Critics of xeno-patents have an argument of their own: patenting genetically engineered pigs blurs the line between human and nonhuman animals in a way that threatens human dignity. The argument is inspired by the work of ethicist David Resnik, who discusses the ethics of patenting human DNA (2001, 2003, 2012). Resnik begins with the observation that human beings have intrinsic moral worth or dignity. The grounds for this dignity may be our rationality, language, sociality, or something else such as the Judeo-Christian notion of being created in "the image of God," but the notion of intrinsic moral worth or dignity explains why human beings have higher moral status than nonhuman animals. It is commonly thought that we matter more than animals, so that if a kitten and child are both drowning and you can save only one, you should save the child—it would be wrong of you to choose to save the kitten over the child, and this is because human beings matter more than kittens. The notion of dignity also explains why it is wrong to exploit others. Slavery treats human beings as mere objects of extrinsic, commercial value, not as beings with intrinsic moral worth. To violate human dignity is to treat "a whole human being, or a part of a human being closely connected to the whole human being, as a complete commodity" (2003, p. 35); to patent a whole human being violates human dignity, because the patent and patent-bearer treat human beings as a commodity, objects of extrinsic value. Violations of human dignity are distinct from threats to human dignity, just as threatening violence is distinct from carrying out violence. A threat to human dignity occurs when "one engages in a practice that could lead to violations of human dignity" (2003, p. 35). To patent human genes, Resnik argues, may threaten human dignity because it would treat a human being as an incomplete commodity. Incomplete commodification can lead to complete commodification, and so, threats to human dignity should be taken seriously.

Do xeno-patents violate human dignity? Presumably not, for xeno-patents do not commodify human beings: it is genetically engineered pigs that are commodified.[2] Do xeno-patents threaten human dignity? While not a topic he takes up explicitly, Resnik may argue that genetically engineered porcine organs are part-human in virtue of having *some* human genetic material. Patenting part-human organs or the process of creating part-human animals treats those part-human parts and animals as commodities, as things to be created and harvested. Since Resnik observes, "partial commodification of human beings can threaten human dignity even if does not violate human dignity," incomplete commodification involved in patenting xenotransplants can promote "exploitation, harm, and injustice, as well as complete commodification of human beings" (2012, p. 276). The history of xenotransplantation research contains instances of exploitation, harm, and injustice, as most xenotransplant recipients are medically fragile, out of therapeutic options, and die shortly after transplantation (recall James Hardy's first attempt at heart transplantation). Even more recently, the case of David Bennett Sr., the recipient of a 2022 cardiac xenograft in the United States, raises questions since he was deemed to be medically noncompliant and thus ineligible for allotransplantation and durable mechanical support. Christopher Gyngell and colleagues (2024) argue that Bennett Sr.'s selection was potentially unfair, for xenotransplant post-monitoring is more invasive than allotransplant post-monitoring, and thus, if medical noncompliance ruled Bennett Sr. out of the latter, then it should have ruled him out of the former. In addition, they raise the concern of whether Bennett Sr., who had no other options, could have made an autonomous choice—how could he have chosen otherwise in his dire circumstance?

The argument regarding xeno-patents violating human dignity is subject to at least three lines of criticism. First, the argument relies on the nebulous claim that human beings have "human dignity." While many would agree that this is true, the notions of "dignity" and "intrinsic worth" have been criticized on grounds that they are vague, imprecise, and "useless" (Macklin, 2003). For instance, if dignity is grounded in our ability to reason, then what about human beings who are unable to reason due to cognitive impairment—do they lack dignity? Different conceptions of human capacities will admit different answers. Second, it is unclear whether xeno-patents even threaten human dignity. The organ that is patented is a pig organ that has been genetically modified to reduce rejection and promote compatibility with the human body. This may involve the insertion of some human genes, but it is hard to see how this would threaten human dignity. Neither human beings nor parts of human beings are being sold or harvested. How it could lead to dignitary harms for human beings is unclear. Finally, as Resnik himself notes, some threats to dignity can be contained, meaning they do not require a wholesale stoppage. To stop would be for society to forego an immense (potential) good. "Some threats are acceptable because their benefits outweigh their harms and preventing those threats would require us to violate basic rights" (2003, p. 35). What is needed, it can be argued, are strict regulation and careful monitoring to ensure that people, especially medically vulnerable people, are not exploited or subject to unethical research. This raises issues of participant selection and informed consent, among others.

Duty to help versus duty to profit

So far, we have neglected to discuss what may be at the forefront of readers' minds: biotechnology companies that create xenotransplants will charge a high price and be guaranteed market share from patent protections. Granting that xeno-patents will help spur the development of clinical xenotransplantation, the patent holders will be interested in making a profit. Indeed, a primary aim of any business is to make money. Milton Friedman describes a corporate executive's role as one of conducting business in accordance with the aims and objectives of the organization's board, and the primary aim of any organization is to "make as much money as possible" (1970, p. 17). This is a discussion about the tension between an organization's social responsibility and a responsibility to profit.

We begin by noting that xenotransplantation is big business. As noted already, corporations and investors have poured tens of millions of dollars into the industry. In March 2021, the biotechnology company eGenesis, which specializes in gene-editing and genome-engineering, announced "the successful completion of a $125 million Series C financing" to advance "the company's proprietary gene-editing" to clinical trials (eGenesis, 2021). More than a dozen healthcare investors contributed, and eGenesis was valued at or above $500 million in 2021. Then, in August 2023, it was announced that Revivicor, a subsidiary of United Therapeutics Corp and the organization that provided the genetically modified pig heart in the first cardiac xenotransplantation, would build a $100 million facility to raise genetically engineered pigs for xenotransplantation (John, 2023). According to one market analysis, the industry is expected to grow from "$13.48 billion in 2023 to $14.58 billion in 2024 at a compound annual growth rate (CAGR) of 8.2%" (The Business Research Company, 2024, np). There is a lot of money being poured into the market, with the expectation of a sizable return on investment.

How much will these companies charge for genetically engineered pig organs? This is a difficult question, given that they are currently not on the market and it is difficult to know. The research is expensive, as one pig-to-baboon cardiac xenotransplant costs around $500,000 (Reardon, 2022). While companies such as eGenesis and Revivicor have not said how much they hope to charge for various organs, there is ample evidence that new medicines and therapies are quite expensive. Novel antiretroviral drugs for HIV were prohibitively expensive, costing around US$15,000–20,000 a year per person. Hoen and colleagues explain that the drugs were so expensive, in part, because they "were largely available only from the originator companies that controlled the patents on these medicines" (2011, p. 1). Patent protections allowed patent owners—that is, pharmaceutical companies—to charge a premium without fear of market competition. In the context of xenotransplantation, David Cooper (2020) has estimated that each genetically engineered pig whose heart, lungs, kidneys, and liver are excised to bring in a quarter of a million US dollars each. It is likely that xenotransplants will cost more than allotransplants, especially in the short term.

No one denies that biotechnology companies that have invested countless dollars and hours into research and that have produced a new source of transplantable organs should seek to profit. It should also be borne in mind that biotechnology companies have an interest in lowering costs: it is good public relations to do so, it aligns with their organizational mission, and a lower price point may result in more sales. But at the end of the day, it is expensive to create xenotransplantable pig organs, and the cost has to be borne by someone. The issue is that charging excessively high prices will result in an unfair distribution of this new source of organs and burden an already burdened healthcare system. The concern is twofold: an unfair distribution of organs and an increase in the burden on healthcare.

Let us begin with equity concerns within the US context—Chapter 8 will discuss global considerations. Xenotransplants will only be available to those who can pay; everyone else is out of luck, meaning they would have to remain on the waitlist for an allotransplant. (This assumes, of course, that private and government insurers will not cover xenotransplantations, an assumption that can be challenged. If xenotransplantation is as viable as allotransplantation, albeit more expensive, it would seem that insurers have reason to cover it. Even if xenotransplants are covered by insurance, out-of-pocket costs would still be high, as evidenced by allotransplantation's significant out-of-pocket costs.) It would seem that only the wealthy will get lifesaving organs, while everyone else is subject to the waitlist lottery. Is it fair for people to potentially miss out on a lifesaving solely because of their financial standing? As socio-economic standing often results from historical injustice, the excessive cost of a xenotransplant may exacerbate existing inequalities.

Defenders of biotechnology companies have a number of responses. First, they can argue that no one has a right to their invention, that just because it is potentially lifesaving does not mean others have a legitimate claim to it. It is their product; they own it; they get to decide what to charge for it. This is not price gouging: they will set a fair market price, one that ensures them good return on their substantial investment. Second, xenotransplants will better everyone's chances of surviving, rich and poor alike. Even if the rich are the only ones who can afford a xenotransplant, that means those who can't afford a xenotransplant have less people to "compete with" on the waitlist. Everyone wins with an influx of new transplantable organs. Third, concerns of exorbitant prices apply to allotransplantation, suggesting that organ transplantation in general is prohibitively expensive for some. A kidney transplant costs more than $400,000, a heart transplant costs around $1.6 million, and a lung transplant costs around $900k to $1.3 million depending on whether it is a single or double lung transplant (Bentley and Ortner, 2020). Ongoing costs for lifelong immunosuppressive therapy can cost thousands of dollars per year (James and Mannon, 2015). Increasingly, potential recipients are being turned away from expensive transplant surgeries due to financial inability to pay the high out-of-pocket costs (Aleccia, 2018). So, concerns about cost barriers are not exclusive to xenotransplantation.

What about the issue of social burden? Being quite expensive, xenotransplants will burden the healthcare system with high costs, especially at the outset. Governments the world over will have a clear interest in lowering costs. This is because many governments offer universal healthcare for citizens, or because—in the US context—the government offers insurance coverage to select vulnerable groups. One proposal would be for the government to enact the "march-in" provision contained in the Bayh-Dole Act (Whalen, 2014). This act, which allows organizations that received federal funding to patent and commercialize inventions, has a provision that allows the government to take control of the invention for certain reasons. More specifically, the march-in provision allows the federal government to require the issuance of a patent license to another party. For example, the government can demand that a biotechnology company issue licenses to competitors for the sake of increasing xenotransplant production and availability, and by doing so, drive down costs. There are four conditions that permit a march-in, the most important for present purposes being that there is a need to alleviate health or safety concerns, concerns that are not being met by the patent holder (Thomas, 2016).

Three responses can be made. First, the assumption that xenotransplantation will increase social costs is not evidently unique to xenotransplantation—any proposal to increase solid organ availability is going to place greater financial pressure on the system in virtue of increasing the number of transplant operations. Heart allotransplantation costs well over $1 million, in addition to lifelong immunosuppression drugs, so increasing the number of heart transplants is going to increase the financial strain on the system no matter how it is accomplished (Bentley and Ortner, 2020). Second, the assumption may well be false, for there is evidence that increasing the number of transplantations may lower overall financial strain. Studies have shown that transplantation is in fact a cost-saving intervention, especially in the context of kidney transplantation as dialysis is an expensive therapy. According to one study, increasing kidney transplantations slightly could lead to "billions of dollars of savings" for the healthcare system over the long term; the authors note, "increases in donation rates of 5% would lead to $4.7 billion in cost savings" (Chen et al., 2021, p. 6). Granting that xenotransplants will be as or more expensive than allotransplants, which is unclear but a reasonable assumption, it is still possible that they will save the healthcare system money overall. Lastly, if the government were to use the march-in provision to require biotechnology companies to issue licenses to competitors in order to lower costs, a primary rationale for patent protection would be undermined: the precedent would be set that biotechnology companies have patent protection in name only, and that if the government decides their pricing is excessive, patent protections would be effectually lifted.

Conclusion

Where does this discussion leave us? With more questions than answers, for sure. There is plenty of hype surrounding xenotransplantation for the future. Moazami and colleagues assert that xenotransplantation could offer an "unlimited, renewable supply of organs independent of human donation" (2023, p. 1989), while Reyneke and colleagues posit that it "holds the potential of a near unlimited supply to solve the ever-growing organ shortage" (2021, p. 2). But how patent protections and the potential for immense profit will translate into clinical reality remains to be argued over: how to ensure equity, access, affordability, and so on in relation to legitimate business ends requires ongoing, honest dialogue. This leads us to our final chapter, which focuses on future directions and global implications.

Notes

1 Some companies obtain patents on inventions adjacent to their market niche to block competitors from entering the market. See Davis and Miller (2000) for discussion.
2 It could be argued that they violate animal dignity, which raises questions about animal moral status and the ethics of raising and killing animals for organs. See Chapter 2.

References

Aleccia, J. 2018. No cash, no heart: Transplant centers need to know you can pay. *The New York Times*, December 4. https://www.nytimes.com/2018/12/04/well/live/organ-transplant-costs-patients.html.

Arora, A., Belenzon, S., Marx, M. and Shvadron, D. 2023. When does patent protection spur cumulative research within firms? *The Journal of Law, Economics, and Organization*, ewad006.

Bentley, T. S. and Ortner, N. J. 2020. *2020 U.S. Organ and Tissue Transplant Cost Estimates and Discussion*. Milliman Research Report. https://www.milliman.com/-/media/milliman/pdfs/articles/2020-us-organ-tissue-transplants.ashx%28

Brody, B. A. 1989. An evaluation of the ethical arguments commonly raised against the patenting of transgenic animals. In *Animal Patents: The Legal, Economic and Social Issues*. London: Palgrave Macmillan UK, 141–153.

The Business Research Company. 2024. *Xenotransplantation Global Market Report*. https://www.thebusinessresearchcompany.com/report/xenotransplantation-global-market-report.

Cartwright-Smith, L. 2014. Patenting genes: What does association for molecular pathology v. myriad genetics mean for genetic testing and research? *Public Health Reports*, 129(3), 289–292.

Chen, H. F., Ali, H., Marrero, W. J., Parikh, N. D., Lavieri, M. S. and Hutton, D. W. 2021. The magnitude of the health and economic impact of increased organ donation on patients with end-stage renal disease. *MDM Policy & Practice*, 6(2), 23814683211063418.

Cooper, D. K. 2019. Financial support for xenotransplantation research. *Xenotransplantation*, 26(3), e12483.

Cooper, D. K. 2020. What will be the cost of a genetically-engineered pig organ for clinical xenotransplantation? *Xenotransplantation*, 27(5), e12606.

Davis, M. H. and Miller, A. R. 2000. *Intellectual Property: Patents, Trademarks, and Copyright in a Nutshell*. West Group.

Diamond v. Chakrabarty, 447 US 303 (1980).

Eberle, M. 1999. March-in rights under the Bayh-Dole act: Public access to federally funded research. *Marquette Intellectual Property Law Review*, 3, 155.

eGenesis. 2021. *eGenesis Announces $125 Million Series C Financing*, March 2. https://egenesisbio.com/press-releases/egenesis-announces-125-million-series-c-financing/.

Fabbri, A., Lai, A., Grundy, Q. and Bero, L. A. 2018. The influence of industry sponsorship on the research agenda: A scoping review. *American Journal of Public Health*, 108(11), e9–e16.

Friedman, M. 1970. The social responsibility of business is to increase its profits. *New York Times Magazine*, 17, September 13.

Goodman, J. R. 2017. Patenting Frankenstein's monster: Exploring the patentability of artificial organ systems and methodologies. *Northwestern Journal of Technology and Intellectual Property*, 15, 35–76.

Greer, T. 2016. UAB awarded $19.5 million grant for new xenotransplantation program. *UAB News*, April 8. https://www.uab.edu/news/people/item/7206-uab-awarded-19-5-million-grant-for-new-xenotransplantation-program

Gyngell, C., Munsie, M., Fujita, M., Thiessen, C., Savulescu, J. and Konstantinov, I. E. 2024. Ethical analysis of the first porcine cardiac xenotransplantation. *Journal of Medical Ethics*, 50(6), 363–367.

Hoen, E. T., Berger, J., Calmy, A. and Moon, S. 2011. Driving a decade of change: HIV/AIDS, patents and access to medicines for all. *Journal of the International AIDS Society*, 14, 1–12.

Holzer, P., Monroy, R. L., Ptitsyn, A., Chang, E., Adkins, J., Brown, T., Rogers, K., Colosi, P. C., Nathwani, A., McIntosh, J. and Tuddenham, E. 2024. Recent patents relating to generating transgenic animals, tissues and cells for xenotransplantation. *Nature Biotechnology*, 42(30), 30.

In Re John Kollar, 286 F.3d 1326 (Fed. Cir. 11 April 2002).

In Re Roslin Inst. (Edinburgh) 750 F 3d 1333, 1337 (Fed Cir. 8 May 2014).

James, A. and Mannon, R. B. 2015. The cost of transplant immunosuppressant therapy: Is this sustainable? *Current Transplantation Reports*, 2, 113–121.

John, S. 2023. Revivicor's $100 million pig-organ facility moves forward. *Virginia Business*, August 30. https://www.virginiabusiness.com/article/revivicors-100-million-pig-organ-facility-moves-forward/.

Joly, Y. 2007. Open source approaches in biotechnology: Utopia revisited. *Maine Law Review*, 59, 382–405.

Kanwar, S. and Evenson, R. 2003. Does intellectual property protection spur technological change? *Oxford Economic Papers*, 55(2), 235–264.

Lever, A. 2008. Is it ethical to patent human genes? In *Intellectual Property and Theories of Justice*. London: Palgrave Macmillan UK, 246–264.

Lim, P. H. and Li, P., 2022. Patentability of biofabricated human organs: Products of nature or products derived from nature revisited. In *Patenting Biotechnical Innovation*. Edward Elgar Publishing, 69–88.

Macklin, R. 2003. Dignity is a useless concept. *BMJ*, 327(7429), 1419–1420.

Moazami, N., Stern, J. M., Khalil, K., Kim, J. I., Narula, N., Mangiola, M., Weldon, E. P., Kagermazova, L., James, L., Lawson, N. and Piper, G. L. 2023. Pig-to-human heart xenotransplantation in two recently deceased human recipients. *Nature Medicine*, 29(8), 1989–1997.

Newman, M., 2023. Johns Hopkins medicine to receive $21.4 million to advance viability of animal organs for transplants and enable human clinical trials. *John Hopkins Medicine Newsroom*, March 29. https://www.hopkinsmedicine.org/news/newsroom/news-releases/2023/03/johns-hopkins-medicine-to-receive-214-million-to-advance-viability-of-animal-organs-for-transplants-and-enable-human-clinical-trials.

O'Sullivan, M. 2018. The legal lacunae of human-animal hybrids and chimeras within patent law. *Journal of Animal Ethics*, 8(1), 62–79.

Reardon, S. 2022. News in focus. *Nature*, 601, 305–306.

Resnik, D. B. 2001. DNA patents and human dignity. *Journal of Law, Medicine & Ethics*, 29(2), 152–165.

Resnik, D. B. 2003. Patents on human-animal chimeras and threats to human dignity. *American Journal of Bioethics*, 3(3), 35–36.

Resnik, D. B. 2012. *Owning the Genome: A Moral Analysis of DNA Patenting*. State University of New York Press.

Reyneke, M., Kerckhof, N., Dherwani, R. and Borry, P. 2021. Should you need an organ . . . Flemish secondary school students' attitudes toward xenotransplantation and transgenetic organ donation. *Xenotransplantation*, 28(6), e12707.

Sellers, M. E. 1994. Patenting nonnaturally occurring, man-made life: A practical look at the economic, environmental, and ethical challenges facing animal patents. *Arkansas Law Review*, 47, 269–297.

Thomas, J. R. 2016. *March-In Rights Under the Bayh-Dole Act*. Congressional Research Service.

United States Constitution. www.law.cornell.edu/constitution

United States Congress. 1987. *New Developments in Biotechnology: Patenting Life*. Congress of the US, Office of Technology Assessment.

US Code. Title 35, Patents. US Code. Title 42, Inventions relating to atomic weapons, and filing of reports.

Whalen, R. 2014. The Bayh-Dole act & public rights in federally funded inventions: Will the agencies ever go marching in. *Northwestern University Law Review*, 109, 1083.

The White House Office of Science and Technology Policy. 2023. *Bold Goals for U.S. Biotechnology and Biomanufacturing Harnessing Research and Development to Further Societal Goals*. https://www.whitehouse.gov/wp-content/uploads/2023/03/Bold-Goals-for-U.S.-Biotechnology-and-Biomanufacturing-Harnessing-Research-and-Development-To-Further-Societal-Goals-FINAL.pdf.

Williams, H. L. 2017. How do patents affect research investments? *Annual Review of Economics*, 9(1), 441–469.

Wouters, O. J., McKee, M. and Luyten, J. 2020. Estimated research and development investment needed to bring a new medicine to market, 2009–2018. *JAMA*, 323(9), 844–853.

8 What are the global and future implications?

Our world is one of inequalities. We see this in numerous areas of life at the international level, from disparate educational access to political instability. One of the most prominent inequalities is in healthcare and access to affordable, effective treatments. This can easily be framed as an issue of global bioethics, as this nascent field examines ethical problems that face persons around the globe related to health, healthcare, related technologies, and policies from nation-states. For instance, a statement from Henk ten Have and Michèle Jean, founders of the field of global bioethics and chief framers of the United Nations Educational, Scientific, and Cultural Organization's Universal Declaration on Bioethics and Human Rights, is emblematic of this point:

> [B]ioethical issues may arise because of inequality and injustice. If an effective medication for diseases such as HIV and AIDS, malaria and tuberculosis is available in some countries, it is morally problematic when patients die in other countries because of a lack of resources.
>
> (2009, p. 23)

This position can easily be reappropriated to xenotransplantation: an effective alternative to allotransplantation may proceed from research into xenotransplantation, and it is likely that this advancement will—due to cost, technology, infrastructure, patent protections, etc.—only be available in certain high-income countries (HICs), leaving other countries behind. There is of course an explanation for this: HICs fund the majority of global health and scientific research, as well as possess the necessary infrastructure, and so those therapies usually become accessible there first. This does not evidently render disparities just, however.

If xenotransplantation demonstrates safety and efficacy as a viable alternative to allotransplantation, and if the ethical and legal challenges surrounding its research and implementation are satisfactorily addressed, it

DOI: 10.4324/9781003472223-9

could have significant global and future implications. While this topic is speculative, some concerns can be anticipated that demand careful consideration. The potential for infectious disease transmission may continue to be a global risk of xenotransplantation, yet this has been extensively discussed elsewhere (see Chapter 3). This concluding chapter will focus on evaluating the potential global impact of xenotransplantation, including (i) global cooperation and governance, (ii) equitable and global accessibility of this biotechnology, (iii) environmental impact, and (iv) some future directions that need additional exploration.

Global governance

There is a need for international oversight of xenotransplant research and implementation. First, the risk and potential benefit are global in nature. If a potential zoonosis resulted from xenotransplant research occurring in Canada, say, then the whole world may suffer as a result, and the suffering will not be uniform: low- or middle-income countries (LMICs) will suffer more than high-income countries (HICs) on account of underfunded or poorly instituted healthcare systems. Moreover, a xenograft recipient may visit or move to a different country, and global cooperation is integral to ensure continuity of care and oversight—nations should know of potential public health threats and recipients would benefit from potential continuity of care. Finally, there is a need for international transparency about how different research teams across the globe are conducting xenotransplant research: how are participants selected and public health protected? What pig lines are researchers in South Korea using and how do these lines compare to those used in Germany and the United States? International oversight would help foster transparency, which is crucial to the advancement of science.

There are established international bodies such as the World Health Organization (WHO) that can set up guidelines and policies for xenotransplant research. But the challenge is that the international bodies lack enforcement mechanisms. Consider climate change, which is a global challenge requiring international cooperation. The Paris Agreement was signed by 196 countries at the United Nations Climate Change Conference (COP21) in 2015: it is supposedly a legally binding document. However, when the United States abruptly withdrew from the Agreement in 2020, there was no repercussion, no way to punish the United States for reneging. Likewise, most countries are unlikely to live up to the Agreement, namely, to meet their goals to cut greenhouse gas emissions. There will be no repercussions for this failure, beyond a warming planet. The point is that international cooperation is difficult not only to implement but to enforce.

Equitable and global access to xenotransplantation

Organ transplantation has long had a problem with equitable access. Equity, defined by the WHO, "is the absence of unfair, avoidable or remediable differences among groups of people, whether those groups are defined socially, economically, demographically, or geographically or by other dimensions of inequality (e.g., sex, gender, ethnicity, disability, or sexual orientation)." Before examining the issue of global access, we should ask: do persons in resource-rich countries like the United States have equitable access to organ transplantation? Let us look at the data. In the United States, black persons develop kidney failure at four times the rate as white persons, yet they are much less likely to receive a kidney transplant (Mullen, 2022). There has been concern voiced that, even in HICs like the United States, xenotransplantation may cause new disparities: "[S]pecifically targeting predominately racial and ethnic minorities for xenotransplantation at the potential risk of inequitable access to human donor transplant (the standard of care) would serve to create a new disparity in transplant care" (Chisholm-Burns et al., 2024). Hence, it has been suggested that xenotransplantation clinical trials must ensure adequate representation from all sectors of the population in order to ensure there are no knowledge gaps regarding safety and efficacy and to build trust among racial and ethnic minorities. This seems like a sound strategy in a high-income context like the United States that is preparing for clinical trials. (Whether it will succeed depends on a lot of unknowns.) Yet, what about the global landscape of xenotransplantation?

If xenotransplantation one day successfully navigates through clinical trials and proves to be an effective therapy, there is a concern that patients from HICs will receive all the benefits while this potentially lifesaving technology would be inevitably slow to reach other areas of the world that are in need. Great disparities already exist in access to transplantation globally. Less than 1% of worldwide transplants are performed in low-income nations annually (White et al., 2014). While the number of transplants in low-income nations is low, it has been hypothesized that in LMICs with transplantation capabilities, the major obstacles to performing more transplantations are (i) logistical elements (e.g., timely transport of organ) rather than infrastructure (e.g., transplant surgeons and patient care abilities) and ii) scarcity of deceased donors (Ward et al., 2018; Awuah et al., 2023). There is a great need for deceased donor transplants in many LMICs as, for instance, much of the global burden of chronic kidney disease occurs there (Stanifer et al., 2016). However, increasing the supply of transplantable organs from deceased donors alone in LMICs may be especially challenging, making alternatives to allotransplantation necessary.

In 2008, the First WHO Global Consultation on Regulatory Requirements for Xenotransplantation Clinical Trials met in Changsha, China.

The summative document that resulted is known as the Changsha Communiqué and it outlines principles on ethics, public health, and key recommendations for the WHO, Member States, and investigators. Despite significant advances in xenotransplantation since its publication, one aspect of the Changsha Communiqué (World Health Organization, 2008) that has been overlooked in recent discussions is Principle 10, which states,

> Because of the potential benefits of successful xenotransplantation, consideration should be given from the beginning to future equitable access to this therapy and the public sector should be encouraged to support xenotransplantation research and development.

Subsequent WHO Global Consultations on Regulatory Requirements for Xenotransplantation Clinical Trials convened in 2011 and 2018 reaffirmed this principle. Of note, the WHO does not prepare these recommendations on its own. The 2011 and 2018 meetings were held in collaboration with the International Xenotransplantation Association—the largest professional society of xenotransplantation academics, researchers, and other interested stakeholders—an official section of The Transplantation Society. Hence, there at least seems to be substantial backing of the recommendations, including Principle 10.

However, it remains undefined what "future equitable access" to xenotransplantation should it prove to be a safe and effective alternative to allotransplantation, means in practice, much less how to attain it. Detailed clarification rather than mere aspiration is needed to address global transplantation disparities, as currently Principle 10 is impracticable without further definition. As discussed in Chapter 7, biotechnology companies will seek to profit from xenotransplants and will enjoy market dominance through patent protections. Who, and with what authority, will encourage these companies to pursue equitable access? What is even meant by *equitable access* is arguably unclear: does it mean that everyone in the world has access or merely many more people than otherwise would have access? Needless to say, more work is necessary.

While a conclusive definition of *equitable access* is elusive, prior WHO publications provide insight. In a 1997 publication on public-private roles in the pharmaceutical sector, the WHO stated that equitable access "means that essential drugs, when needed, are available to and affordable by all" (World Health Organization, 1997). This aspiration for universal access is unrealized with essential medicines. The WHO maintains country-specific model lists of medicines considered to be the important needs of a particular health system. The WHO estimates that around two billion people globally lack access to essential medicines due to unavailability and unaffordability, depriving them of the significant advancements and benefits of modern science and medicine. Administering essential medicines, many delivered by mouth, is exceedingly more straightforward to provide a patient than a transplantation, let alone a genetically modified pig-to-human transplantation. Therefore, achieving the standard envisioned by WHO for xenotransplantation poses significant challenges.

While xenotransplantation as a clinical alternative to allotransplantation is years—maybe even decades—away (if ever), there is a need for more discussion on equitable access if indeed the WHO and international xenotransplant community are committed to ensuring that not only high-income countries have access to this medical advance. Progress is possible, even now, on this. Although the Changsha Communiqué is not legally binding upon Member States or researchers, its ethical influence, owing to the WHO's global prominence, holds significant weight. Currently in the United States, biotechnology companies predominantly fund xenotransplantation research and hold patents for the genetically modified pigs. Hence, any vision of future equitable access globally must involve these companies. Having the WHO initiate conversations with these companies and stakeholders, including Member States, is a crucial first step toward equitable access. Biotechnology companies can be urged to integrate equitable access into their corporate social responsibility strategy. Topics such as pricing regulations, subsidies for those in need, scaling up production of genetically modified pigs to meet the shortage of organs, schemes for opening markets in LMICs so that xenografts can be procured near the transplant site, and associated costs require thoughtful deliberation.

One potential area of research that could promote access to xenotransplantation globally is the advancement of cryopreservation—a process in which biological materials (e.g., tissues and organs) are stored at extremely low temperatures, usually in liquid nitrogen, to preserve them for later use (Giwa et al., 2017). There is a current consensus that it is possible to store tissues and organs for extended periods at very low temperatures—low enough to nearly cease cellular activities (less than –130°). In practice, however, significant challenges remain, particularly for complex vascularized organs, such as the formation of ice crystals and the correct protocol for cooling and rewarming the biological material. While cryopreservation has been applied in some clinical settings for certain types of cells, such as sperm and embryos, scaling these techniques to whole organs—especially large and functionally complex ones like the heart, lungs, and kidneys—remains an ongoing scientific challenge (Chen et al., 2023).

Nonetheless, if these technical challenges can be overcome, cryopreservation could play a transformative role in the future of xenotransplantation. Currently, xenotransplantation is being undertaken in contexts relatively close to the pig facility that raises the genetically modified pig. The rate-limiting step in this organ transplant model is having a highly controlled, biosecure pig facility relatively close by. Viable organs must be transplanted soon after procurement, necessitating proximity between the pig facility and the recipient. This requirement restricts the feasibility of xenotransplantation to locations with both the necessary infrastructure and a nearby source of organs. Yet, what if a pig organ could be procured from a donor animal in the United States and then, through cryopreservation, it could be shipped to the other side of the globe and potentially even stored for future use? This would enable xenotransplantation in regions without direct access to donor pig facilities. Such advances could not only mitigate the current logistical constraints but also allow for the establishment of "xenograft biobanks" where standardized, quality-controlled organs are stored and made available on demand.

Researchers have hypothesized that the ability to bank inventories of xenografts could make xenotransplantation more practical and cost-effective at scale (Giwa et al., 2017). Of course, the ethical and regulatory implications of such a system would also need to be considered in the future. Questions regarding equitable distribution, international regulations, and the potential for commercialized xenograft markets would arise, necessitating systems for oversight and ethical frameworks. Still, the ability to cryopreserve xenografts could mark a shift in transplantation medicine, addressing not only the global organ shortage but also the infrastructural disparities that currently limit access to lifesaving treatments.

Lastly, an aspect of global xenotransplantation—one that, to our knowledge, is not currently under consideration but has a storied history within the pharmaceutical industry—is outsourcing clinical trials to low- or middle-income countries (LMICs) (Petryna, 2009). This occurs for a number of reasons, including that it is often less expensive to conduct research in LMICs rather than HICs, research participants are plentiful, and there is—at times—more lax oversight. The ethical problem is that research is being done on a population that will not benefit from that research—at least not in the near term. Because of this, certain ethics guidelines have introduced aspects of benefit-sharing or post-study obligations that researchers (or, more specifically, the funders of research) should follow (Hurst, 2017). While, again, we are not aware that researchers are proposing this methodology for xenotransplantation and would be doubtful that it would occur given the great medical and technological elements that transplant requires, this would not be far-fetched. Some middle-income countries such as Thailand have established transplant systems and may be open to a biotechnology company to conduct xenotransplant research. Given how much financial investment has been poured into the research already, if the US Food and Drug Administration is perceived to be slow in approving xenotransplant clinical research, biotechnology companies would have a clear incentive to pursue research elsewhere. This would be an area in need of further analysis should it be proposed.

Environmental impact

One claim raised by critics of xenotransplantation is that if it proceeds to demonstrate safety and significant clinical efficacy, then it would result in saving millions of people's lives who would have otherwise died; consequently, this would lead to an unprecedented increase in the population, which—it is claimed—would wreak havoc on the environment with the resulting harm outweighing the benefits of xenotransplantation (Bisong and Asuo, 2020). That is, as the argument goes, xenotransplantation has the possibility of saving too many lives, significantly increasing the population, and causing further harm to the environment—curing disease would pose an existential risk.

However, if we take this augment to its conclusion, we would have to surmise that we should stop trying to cure other serious diseases such as stroke, diabetes, and dementia. Xenotransplantation should not be singled out, for the concern of saving too many lives applies to many if not all therapies. Furthermore, this argument is predicated on the most optimistic account of xenotransplantation, whereby the outcomes of a xenotransplant are equal or superior to an allotransplant, and that everyone who needs an organ transplant can feasibly receive one. However, this belies how clinically realistic it is to ensure that everyone who needs a transplant *can* receive one. One obvious problem is that there are just not enough transplant surgeons to perform thousands of additional organ transplants each year. The argument also presumes that xenotransplantation would primarily be used as a destination therapy rather than a bridge until a human organ becomes available: it is unlikely to be so effective, at least in the very near future. As a bridge therapy, it would not lead to a cataclysmic population increase. The argument also ignores other mitigating factors such as the current global trajectory in fertility rates. For example, the global population is unlikely to continue to increase throughout the century; 23 countries—including Spain, Japan, and Thailand—are expected to see their populations halve by 2100 (Vollset et al., 2020). For these reasons, the claim that permitting xenotransplantation would lead to a catastrophic environmental disaster lacks plausibility.

Others have made the baseless claim that using animals as a source of xenografts would "make most animals to go into extinction and thereby reducing the already over depleted biodiversity in the world" (Bisong, 2015). Furthermore, it has been argued that the potential of xenozoonosis is "just like a time bomb waiting to explode. If it does lead to a world plague, who knows, perhaps, the whole world would be wiped out" (Bisong, 2015). Therefore, as argued by this particular commentator, xenotransplantation research should be halted and not permitted to progress to clinical trials. Yet, while the most recent research continues to hold that unknowns remain with regard to xenozoonotic infection, the risk is thought to be low and manageable (see Chapter 3). Second, it is, to our minds, indefensible, that xenotransplantation will cause species extinction; no published data modeling shows this to be true, and in the absence of evidence for such a bold assertion, it can be dismissed.

To date, there have not been studies assessing what might be the environmental impact, in terms of climate change, of xenotransplantation. The raising of pigs for meat accounts for nearly 750 million tons of CO2e greenhouse gas (GHG) emissions annually across the globe (Zhang et al., 2024). While raising pigs for xenotransplantation will constitute a small fraction of GHG, it is plausible that there will be some net increase in GHG emissions due to additional pigs being raised: for example, food will have to be grown, harvested, and transported to the facilities, and there will be increased emissions from transporting the pigs or their organs to transplant centers, not to mention potentially more emissions resulting from an increase in the number of transplant recipients. In the future, as part of being a socially responsible company, this may merit further assessment by the companies involved in producing genetically engineered pigs for xenotransplantation and seeking ways to offset it.

Future directions

There are several areas of xenotransplantation that need additional exploration in the future as the world begins clinical trials and the possibility of xenotransplantation becomes clinical reality.

Global registry

As we write this in early 2025, the United States has now approved kidney xenotransplantation clinical trials. In addition, China has performed xenotransplantation in decedents and liver xenotransplantation in a living patient, and it has been reported that Russia, Japan, Germany, and likely other countries are also forging ahead. As countries move ahead with their respective programs, there is a need for information sharing and coordination of efforts. The 2008 Changsha Communiqué states the following:

> There needs to be a global system for exchanging information, preventing unregulated xenotransplantation, providing support for states and coordinating xenotransplantation vigilance, surveillance and response to suspected infections.

The rationale for this is straightforward: a global registry would facilitate transparency and thereby help protect public health.

At present, if an individual wants to know about current research in xenotransplantation, they have a few options at their disposal. Some countries have national clinical trial registries, such as the one compiled by the United States Department of Health and Human Services (www.clinicaltrials.gov). There is also a database set up by the World Health Organization, International Xenotransplantation Association, and Sichuan (China) Provincial People's Hospital that monitors certain aspects of xenotransplantation studies (www.humanxenotransplant.org). David Cooper (2020) has argued that, while these databases provide a starting point, a more concerted effort is needed so that the experience gained from clinical trials and other research studies (such as decedent studies and expanded use) is readily available to interested researchers. Cooper recommends that, at a minimum, the data that needs to be captured in an accessible database would include institutional data (e.g., the institution and researchers), patient data, pig data, and transplant and outcome data. In addition to this list, we could add that sharing of best practices regarding informed consent and monitoring of patients for xenozoonosis is of high importance. In the expanded access uses of xenotransplantation that have been completed in the United States to date, researchers have shared very little information about these aspects publicly, and it is uncertain if they have done so privately with other researchers in the field. Yet, for the field to progress, sharing of this information seems essential.

Care for the body

In the 2021 drama *The God Committee,* a hospital organ transplant committee has just one hour to deliberate and decide which of the three patients should receive a heart transplant. Factors such as the patients' social support systems, disposition, and mental health are raised in the deliberation. One patient is classified as obese, which some committee members believe precludes him from receiving the heart. At present, healthy organs for transplant are scarce, yet this will not necessarily be the case should xenotransplantation prove to be a safe and effective clinical alternative to allotransplantation. There may be a day when hospital transplant committees no longer regularly deliberate on who should receive the organ because organs are plentiful. If such a day arrives, values regarding care for the body and health may come to the fore.

If xenotransplantation becomes a clinical reality, it may alter how we think about the value of our corporeal being. For instance, why should I care about my heart health if there are organs readily available? Should I worry about heart disease if I know that spare parts are available that can "fix" me? Should I worry about not drinking too much alcohol so as to avoid kidney and liver disease? While speculative, xenotransplantation may affect our outlook on bodily integrity and health, and these are questions that may need exploring in depth in the future.

Recipient stigma

In allotransplantation, Sharp (1995) discusses stigmas around organ recipients: others view them as not fully able-bodied but as disabled. At the same time, Sanner (2001) discusses the importance of the origin of the organ to potential recipients: it often matters whose organ one receives. These stigmas and psychological concerns are likely to be exacerbated, and indeed, numerous studies of xenotransplant perception indicate worries about receiving a genetically engineered porcine organ. In a recent study of attitudes in the United States, Padilla et al. found that around 40% of people were hesitant (Padilla et al., 2024). More than 30% of parents expressed concern over or refusal of a pediatric xenotransplant (Padilla et al., 2021). Several parents of children with congenital heart disease voiced concerns that their child may encounter bullying due to having a pig heart (Hurst et al., 2021). Such studies indicate broad concerns based on psychological, religious, and societal grounds and support the possibility that stigma could become associated with receiving a genetically engineered porcine organ and that recipients may experience higher rates of anxiety and stress as they navigate a cultural taboo.

Business interests

If xenotransplantation's promises are to be believed, it would radically alter the existing organ donation systems the world over by offering a business model solution. Rather than relying on altruistic donations, transplantable organs will be available for purchase from a biotechnology company. Whereas organ buying and selling is strictly forbidden in the United States and many other countries (Iran aside), introducing purchasable xenografts generates financial incentives that have thus far been resisted. As mentioned in the previous chapter, this may introduce powerful lobbying against public health and preventative measures that promise to reduce demand for organ transplantation. The introduction of a for-profit business model for transplantation incentivizes transplantation, not preventative measures that lower transplant demand. So, it may generate perverse incentives to the contrary (Bobier et al., 2025).

Currently, at least in our United States context, dialysis centers are a big business. Approximately 80% of dialysis centers in the United States are operated as for-profit entities. As a whole, the dialysis industry in the United States treats over 400,000 patients annually, with Medicare reimbursements amounting to around US$33 billion each year (Eliason et al.,

2019). Given the heavy reliance on government-funded healthcare reimbursement, dialysis has become an entrenched and lucrative sector of the medical economy. However, a future in which kidney xenotransplantation is an effective and widely available therapy for end-stage renal disease could significantly disrupt this economic landscape. Unlike dialysis, which requires ongoing and costly treatments multiple times per week, a successful xenotransplant could offer a long-term, potentially curative solution. Xenotransplantation could substantially reduce the demand for dialysis services, leading to a marked decline in patient volume at dialysis centers. This shift would likely have major financial implications for the for-profit dialysis industry. It remains to be seen how the dialysis industry may respond to these threats on its business interests.

Within allotransplantation, current donation-based systems incentivize public health measures, for there are insufficient transplantable organs, and societies need to lower their demand. By introducing business incentives into the transplant system, there will be reasons to maintain the status quo and resist measures that threaten profitability, that is, lower transplant demand. There will also be new interests in protecting the profitability of xenografts, and hence, reasons to resist international efforts to lower cost.

There are well-documented and compelling examples of pharmaceutical companies lobbying against public health measures that threaten their business interests. Wouters found that the pharmaceutical and health industry spends an average of around $233 million per year in the United States on federal lobbying and that "the industry focused its efforts on opposing major drug cost-containment measures by contributing to ballot measure committees in key states" (2020, p. 693). For example, the pharmaceutical industry spent hundreds of millions of US dollars to prevent the passage of the Inflation Reduction Act of 2022, which would have allowed the Centers for Medicare & Medicaid Services to negotiate drug prices, regulate insulin prices, and cap out-of-pocket spending for Medicare recipients, all of which would promise to reduce national health spending and help individuals afford healthcare (Sayki, 2023).

Allotransplant waitlist

What may become of allotransplantation if xenotransplantation becomes a safe, effective, and accessible therapy? The demand for allotransplantation could be significantly reduced as patients who would otherwise rely on human donor organs might instead receive xenografts, potentially shifting organ transplantation from a scarcity-driven model to one based on clinical suitability and patient preference. It is foreseeable that this shift could lead to a decline in living and deceased organ donation. At the same time, ethical, religious, and cultural objections to xenotransplantation could sustain a niche demand for allotransplantation among certain patient populations.

Despite these changes, we do not think that allotransplantation will disappear entirely. Some patients may be better suited for human donor organs due to immunological compatibility or medical factors, and allotransplantation may still offer advantages in specific cases. Ultimately, at least in the near future, rather than replacing allotransplantation altogether, xenotransplantation is more likely to coexist with it.

Public engagement

There has been a rise in empirically informed bioethics since the turn of the 21st century. While there remains contention of *how* the empirical should inform the normative—an argument going back to at least David Hume's famous "is-ought problem"—many accept that the empirical should *somehow* inform how ethics and policy proceed. Engaging the public in discussion to understand their viewpoints, any misconceptions that are held, and any hesitations is a needed step when introducing a novel medical product.

The importance of community engagement is recognized in international research ethics guidelines, such as the United Nations Educational, Scientific, and Cultural Organization's "Universal Declaration on Bioethics and Human Rights," as well as in the "International Ethical Guidelines for Health-related Research Involving Humans," a guidance document jointly published in 2016 by CIOMS and the WHO, which states the following:

> Guideline 7: Researchers, sponsors, health authorities and relevant institutions should engage potential participants and communities in a meaningful participatory process that involves them in an early and sustained manner in the design, development, implementation, design of the informed consent process and monitoring of research, and in the dissemination of its results.

Within xenotransplantation, the importance of public engagement has been recognized. The original Changsha Communiqué (2008) specifies that any xenotransplantation regulatory system "should be transparent, must include scientific and ethical assessment and should involve the public." The WHO, in concert with the International Xenotransplantation Association, reaffirmed this principle in its subsequent versions in 2011 and 2018.

Public engagement involving a diverse set of perspectives may help ensure that cultural, religious, socio-economic, and demographic considerations are taken into account as xenotransplantation moves from the lab to the clinic. This has been the recommendation of several commentators. For instance, Peter Reese and colleagues (2024) have recently advocated that the biotechnology companies that are actively involved in bringing xenotransplantation to the clinic should establish patient advocacy boards to highlight the concerns of individuals with end-stage kidney disease.

Further, these companies should proactively engage with communities of racial and ethnic minorities to inform them about xenotransplantation and address their concerns. These seem to be needed steps in the area of public engagement around xenotransplantation. Yet, while ethics guidelines and documents such as the Changsha Communiqué espouse the importance of public engagement, we must recognize that public engagement has not been done systematically. There is, to date, only one very recent study of widespread viewpoints across the United States (Padilla et al., 2024), with a few other single-center studies. Globally, the empirical data is similarly sparse. Published data on viewpoints in many countries are either scant or nonexistent.

Public engagement should involve both qualitative and quantitative research methods (Hurst and Cooper, 2024). These approaches can complement each other, with qualitative research serving as a precursor to broader quantitative studies. The qualitative research should target specific communities surrounding the institution where the xenotransplantation clinical trials are expected to occur. This research could include various methods such as one-on-one interviews and focus groups with key stakeholders. These stakeholders should include patient groups, such as those needing transplants and individuals who have already undergone allotransplantation, as well as influential community members like policymakers and religious leaders. It is essential to ensure diversity within these groups, representing a wide range of demographics, genders, and ages, with a deliberate effort to include both young persons and seniors. Differences in attitudes toward xenotransplantation have been found based on race (Padilla et al., 2020) and religion. Additional studies should be conducted to confirm these findings and consider if there are ways to reduce hesitations. Based on the findings from these qualitative studies, broader quantitative studies can then be conducted to gauge wider public opinion.

Conclusion

As we have explored throughout this volume, xenotransplantation presents myriad fascinating ethical questions that are necessary to explore at the local and global levels. The field is moving fast. At the time of this writing, only two early clinical trials exist in the world for kidney xenotransplantation. Yet, this may quickly change. While the field is moving rapidly, many of these ethical issues will continue. Because of these enduring issues, we have attempted to give the reader a broad overview of the current ethics debate in the field and what is on the near horizon. We end with our invitation to join the ongoing important conversations surrounding this groundbreaking technology.

More work is yet to be done. Empirical viewpoints of persons toward xenotransplantation have only begun to be studied systematically in the United States—the worldwide leader in xenotransplantation at this point. It will be necessary for researchers to study viewpoints in order to understand attitudes and where hesitations may exist among the population. Further, we have barely touched on the topic of healthcare expenditure. It is plausible that, should xenotransplantation become a clinical reality and adopted by many countries as standard practice, people may live longer and, thereby, require healthcare for a longer period than otherwise. This will have an impact on healthcare budgets. As many questions remain unsettled, and as clinical trials begin, we are confident that new concerns will present themselves. The need for ethicists, social science researchers, and others to be involved from the outset in forming the normative ethical guidelines to guide this burgeoning field is paramount. We are hopeful that this volume, in some small way, helps to inform this conversation.

References

Awuah, W. A., Ng, J. C., Bulut, H. I., Nazir, A., Tenkorang, P. O., Yarlagadda, R., Mustapha, M. J., Abdul-Rahman, T., Akilimali, A., Oti, V. N. and Aji, N. 2023. The unmet need of organ transplantation in Africa. *International Journal of Surgery*, 109, 519–520.

Bisong, P. B. 2015. Interference with nature: Xenotransplantation procedure and its potential effects on man. *Journal of Health Ethics*, 11, 2.

Bisong, P. B. and Asuo, O. 2020. Xenotransplantation technology and its potential effects on the environment. *International Journal of Environmental Pollution and Environmental Modelling*, 3, 121–128.

Bobier, C., Gibson, R. B., Merlocco, A., Rodger, D. and Hurst, D. J. 2025. Xenotransplantation as a business solution to the organ shortage. *Bioethics*, 39(5), 503–511.

Chen, J., Liu, X., Hu, Y., Chen, X., & Tan, S. 2023. Cryopreservation of tissues and organs: Present, bottlenecks, and future. *Frontiers in Veterinary Science*, 10, 1201794.

Chisholm-Burns, M., Kelly, B. S. and Spivey, C. A. 2024. Xenotransplantation could either be a friend or foe of healthcare equity. *Communications Medicine*, 4, 85.

Cooper, D. K. C. 2020. Clinical trials of xenotransplantation: The need for a worldwide registry. *Xenotransplantation*, 27, e12598.

Council for International Organizations of Medical Sciences. *International Ethical Guidelines for Health-Related Research Involving Humans*. https://cioms.ch/wp-content/uploads/2017/01/WEB-CIOMS-EthicalGuidelines.pdf (Accessed: 26 August 2024).

Eliason, P. J. et al. 2019. How acquisitions affect firm behavior and performance: Evidence from the dialysis industry. *The Quarterly Journal of Economics*, 135(1), 221–267.

Giwa, S. et al. 2017. The promise of organ and tissue preservation to transform medicine. *Nature Biotechnology*, 35, 530–542.

Hurst, D. J. 2017. Benefit sharing in a global context: Working towards solutions for implementation. *Developing World Bioethics*, 17, 70–76.

Hurst, D. J. and Cooper, D. K. C. 2024. The importance of public engagement in clinical xenotransplantation. *Health Care Science*, 3, 124–130.

Hurst, D. J., Padilla, L. A., Cooper, D. K. C. & Paris, W. 2021. Factors influencing attitudes toward xenotransplantation clinical trials: A report of focus group studies. *Xenotransplantation*, e12684.

Mullen, J. 2022. *How Our Organ Transplant System Fails People of Color*. AAMC. https://www.aamc.org/news/how-our-organ-transplant-system-fails-people-color#:~:text=In%20the%20United%20States%2C%20Black,rates%20than%20their%20White%20counterparts (Accessed: 26 August 2024).

Padilla, L. A., Hurst, D. J., Jang, K., Rosales, J. R., Sorabella, R. A., Cleveland, D. C., Dabal, R. J., Cooper, D. K., Carlo, W. F. and Paris, W. 2020. Racial differences in attitudes to clinical pig organ xenotransplantation. *Xenotransplantation*, e12656.

Padilla, L. A., Hurst, D. J., Zink, A., Parent, B. and Kimberly, L. L. 2024. Public attitudes to xenotransplantation: A national survey in the United States. *American Journal of Transplantation*, 24(11), 2066–2079.

Padilla, L. A., Rhodes, L., Sorabella, R. A., Hurst, D. J., Cleveland, D. C., Dabal, R. J., Cooper, D. K., Paris, W. and Carlo, W. F. 2021. Attitudes toward xenotransplantation: A survey of parents and pediatric cardiac providers. *Pediatric Transplantation*, 25, e13851.

Petryna, A. 2009. *When Experiments Travel: Clinical Trials and the Global Search for Human Subjects*. Princeton, NJ: Princeton University Press.

Reese, P. P., Powe, N. R. and Lo, B. 2024. Engineering equity into the promise of xenotransplantation. *American Journal of Kidney Diseases*, 83, 677–683.

Sanner, M. A. 2001. People's feelings and ideas about receiving transplants of different origins—questions of life and death, identity, and nature's border. *Clinical Transplantation*, 15, 19–27.

Sayki, I. 2023. Despite record federal lobbying spending, the pharmaceutical and health product industry lost their biggest legislative bet in 2022. *Open Secrets*, February 2. https://www.opensecrets.org/news/2023/02/despite-record-federal-lobbying-spending-the-pharmaceutical-and-health-product-industry-lost-their-biggest-legislative-bet-in-2022/.

Sharp, L. A. 1995. Organ transplantation as a transformative experience: Anthropological insights into the restructuring of the self. *Medical Anthropology Quarterly*, 9, 357–89.

Stanifer, J. W., Muiru, A., Jafar, T. H. and Patel, U. D. 2016. Chronic kidney disease in low- and middle-income countries. *Nephrology Dialysis Transplantation*, 31, 868–874.

ten Have, H. A. M. J. and Jean, M. S. (2009) Introduction. In ten Have, H. A. M. J. and Jean, M. S. (eds.), *The UNESCO Universal Declaration on Bioethics and Human Rights: Background, Principles and Application*. Paris: UNESCO Publishing, 17–55.

Vollset, S. E., Goren, E., Yuan, C.-W., Cao, J., Smith, A. E., Hsiao, T., Bisignano, C., Azhar, G. S., Castro, E., Chalek, J., Dolgert, A. J., Frank, T., Fukutaki, K., Hay, S. I., Lozano, R., Mokdad, A. H., Nandakumar, V., Pierce, M., Pletcher, M., Robalik, T., Steuben, K. M., Wunrow, H. Y., Zlavog, B. S. and Murray, C. J. L. 2020. Fertility, mortality, migration, and population scenarios for 195 countries and territories from 2017 to 2100: A forecasting analysis for the global burden of disease study. *The Lancet*, 396, 1285–1306.

Ward, A., Klassen, D. K., Franz, K. M., Giwa, S. and Lewis, J. K. 2018. Social, economic, and policy implications of organ preservation advances. *Current Opinion in Organ Transplantation*, 23, 336–346.

White, S. L., Hirth, R., Mahillo, B., Dominguez-Gil, B., Delmonico, F. L., Noel, L., Chapman, J., Matesanz, R., Carmona, M., Alvarez, M., Nunez, J. R. and Leichtman, A. 2014. The global diffusion of organ transplantation: Trends, drivers and policy implications. *Bulletin of the World Health Organization*, 92, 826–835.

World Health Organization. 1997. *Action Programme on Essential Drugs. Public-Private Roles in the Pharmaceutical Sector: Implications for Equitable Access and Rational Drug Use.* https://apps.who.int/iris/handle/10665/63693 (Accessed: 26 August 2024).

World Health Organization. 2008. First WHO global consultation on regulatory requirements for xenotransplantation clinical trials: Changsha, China, 19–21 November 2008. The Changsha communiqué. *Xenotransplantation*, 16, 61–63.

World Health Organization. *Health Equity*. https://www.who.int/health-topics/health-equity#tab=tab_1 (Accessed: 26 August 2024).

Wouters, O. J. 2020. Lobbying expenditures and campaign contributions by the pharmaceutical and health product industry in the United States, 1999–2018. *JAMA Internal Medicine*, 180, 688–697.

Zhang, L., Mao, Y., Chen, Z., Hu, X., Wang, C., Lu, C. and Wang, L. 2024. A systematic review of life-cycle GHG emissions from intensive pig farming: Accounting and mitigation. *Science of the Total Environment*, 907, 168112.

Index

Note: Page numbers in *italics* indicate a figure on the corresponding page.

3D bioprinting 21

Agrawal, M. 68
allotransplantation: costs 119–121; demand and supply issues 10–14, 31–32, 134–135; as gold standard 10, 19, 70; heart 120–121; pediatric 92–93, 98–99; stigma 133; in US 11–12, 18–19, 126; waitlist mortality 13, *14*, 62–63, 65, 92–93, 134–135; vs. xenotransplantation 1, 19–21, 69–72, 98, 135
alternatives to xenotransplantation: 3D bioprinting 21; artificial liver systems 71; bridge therapy 13; destination therapy 13, 63, 71–72; dialysis 13; ECMO 13; expanded criteria donation 17; ex vivo machine perfusion 17; Impellas 13; lab-grown organs 21, 31–32; left ventricular assist device (LVAD) 1–2, 13; lifestyle measures 12, 31; living donation expansion 15; machine perfusion methods 17; mechanical organs 21; organ markets 15, 32; preventative medicine 31; regenerative medicine techniques 21; synthetic organs 21–22; total artificial hearts 13
American Medical Association, *Code of Medical Ethics* 61
animal moral status 26; modest position 30; strong position 27–28; weak position 28–29
animal research, ethical principles 30; basic needs 36; cost-benefit analysis 32–34; no alternative method 31–32; no unnecessary harm 35–36; sufficient value to justify harm 34–35; upper limits to harm 36–37
animal studies 33
Animal Welfare Act 114
Arshad, A. 32
artificial liver systems 71
authority-retaining consent 53

Baby Fae experiment (Stephanie Fae Beauclair) 4, 67–68, 91–92, 98; clinical trials 93–97; informed consent issues 100–101; pediatric considerations 92–93; psychosocial impact 99–100; public reaction 97–99
Bailey, L. 91
Barnard, C. 3
Bayh-Dole Act 115, 119
Beauchamp, T. L. 30–36
Belmont Report 65, 69, 76–78, 94
benefits 47, 49, 59–60, 66–69, 95; of animal research 32–34; informed consent 85–86
Bennett, D. 116
biosurveillance 78, 81, 83, 87–88, 100; long-term 50–54
brain death *14*, 15–16
bridge therapy *see* destination therapy
Brody, B. 114
business interests 112, 133–134

Caplan, A. 67
case study *see* Baby Fae experiment (Stephanie Fae Beauclair)

Centers for Disease Control and Prevention (CDC) 84
Centers for Medicare & Medicaid Services 134
Changsha Communiqué 61, 127–128, 131–132, 135–136
circulatory determination of death (DCDD) 16–17; *see also* organ donors
climate change 125, 131
Cochrane, A. 28
Code of Federal Regulations 78
collective consent 5, 46–50; *see also* informed consent
Commerce Clause (US Constitution) implications 83–84
contractarianism 29
Cooper, D. 12, 62–63, 65, 70–72, 83, 114–115, 118, 132
corporate social responsibility strategies 128
costs 118–120; healthcare 137
Council for International Organizations of Medical Science (CIOMS) 94–95, 135
COVID-19 pandemic 13, 17, 66
CRISPR gene-editing technology 107
cryopreservation technology 128

Daar, A. S. 82
DeGrazia, D. 30–36
Denner, J. 45
destination therapy 13, 63, 71–72, 95, 130
dialysis industry disruption 133–134
Diamond v. Chakrabarty 107–108

eGenesis 117–118
Emanuel, E. J. 59, 67–68, 79–80, 85–86
employment-like contracts 53–54; *see also* Ulysses contract
environmental impact 130–131
equitable access 126–130
equity, defined 126
ethics of xenotransplantation 25–30, 94–97, 106–110; consequentialist arguments 110–114; non-consequentialist arguments 114–117; *see also* animal research, ethical principles
expanded access criteria 84–85, 96–97

fairness of limiting trials 60–64
fair subject selection 59–60; fair opportunity 66; generalizability 64–65; shared burden 65–66; third-party risk 66–67
Fenton, K. 5
fertility rate trends 130
Food and Drug Administration (FDA) 3, 33, 61, 63, 84, 97
Fovargue, S. 44, 46
Fox, R. R. C. 70
Francione, G. L. 27
Friedman, M. 117

genetic engineering 37–38
global access 126–131
global governance 125
global registry 131–132
The God Committee (drama) 132
graft rejection 1, 12, 22, 86, 93–94, 96, 108
Gyngell, C. 116

Hardy, J. 3, 75
Harvard Oncomouse patent 108
Have, H. 124
heart transplants 3, 75, 78, 92–98, 116, 120
high-income countries (HICs) 124–126, 129
Hoen, E. T. 118
health attitudes 132–133
human-decedent studies 33
human dignity 115–117
human research ethics 59–60
Hume, D. 135

immunosuppression 21, 64, 87, 99–100, 120
inequality 20, 119, 124–128; high-income countries (HICs) 124–126; low- and middle-income countries (LMICs) 125–126
Inflation Reduction Act of 2022 134
informed consent 60, 69, 75–77, 87–88; and lifelong monitoring 100–101; media attention 84–85; nonadherence 86–87; potential restrictions 82–83; quarantine 83–84; recommendations 87–88; risk-benefit ratio 85–86; standards for 77–78; undue inducement 78–80; withdraw issues 81–82

International Ethical Guidelines for Health-related Research Involving Humans 135
International Xenotransplantation Association 127

Jean, M. 124
Johnson, L. S. M. 34, 44, 47
Jorqui-Azofra, M. 36, 46

Kabasenche, W. P. 38
kidney (xeno)transplantation 22, 32–34, 79, 96, 136; costs 119–120, 134; ethics of 25; health deterioration 11; in UK 15; unequal access 126; waiting lists 13, 63
Knoll, E. 68
Kong, W. M. 86

Lanza, R. 12, 65, 70–72
left-ventricular-assist device (LVAD) 1–2, 13, 71
lifelong monitoring 100–101
Li, P. 108
Lim, P. H. 108
lifestyle measures 12
limiting trials: fairness 60–64; safe alternative 61–62
living donation 13–15, *14*, 20
Łopata, K. 43
low and middle-income countries (LMICs) 125–126, 128–129
Lundberg, G. D. 68
Lynch, H. F. 53

MacKay, D. 64–66, 86
march-in provision 119–120
Mary Mallon case 42
McConnell, T. 52
media attention 84–85
Montgomery, R. A. 33
Mukherjee, R. 43
Murphy, K. N. 38
Murray, J. E. 20

National Institutes of Health 114
nonobvious intervention 107–109
Norwood procedure 91
Nuffield Council on Bioethics 83, 93–94

organ donation: euthanasia 16; systems 32, 133–134
organ donors 13–18, *14*, 20, 126; dead donor rule 6, 16–17; "opt out" *vs* "opt in" policies 6, 17, 32
organ shortage 10–13, 31, 47, 92–93, 120, 129; *see also* inequality
organs supply and demand 13, 15–20, 133–134
Ost, S. 44
O'Sullivan, M. 108, 114

Padilla, L. A. 87
pain modification 37–38
Paris Agreement 125
Patent and Trademark Act Amendments of 1980 115
Patent and Trademark Office (PTO) 108–109
patents 105–106; *see also* xenotransplant patents
pathogenic microorganism potential 81–82; *see also* zoonotic infection
pediatric xenotransplantation: clinical trials 93–97; lifelong monitoring 100–101; need for 92–93; psychosocial considerations 99–100; public view 97–99
pharmaceutical industry lobbying 133–134
Pierson III, R. N. 62
pig(s): genome 42–43, 45; environmental impact of 131; with human genes 109; pain modification 37–38; pig-to-baboon studies 85, 96, 118; pig-to-human xenotransplants 3–5, 128; public health concerns 6–7; raising facilities for 128–129; well-being *see* animal research, ethical principles; *see also* animal moral status; zoonotic infection
porcine cytomegalovirus (PCMV) 43, 45–46; *see also* zoonotic infection
porcine endogenous retroviruses (PERVs) 43, 45–46; *see also* zoonotic infection
posttransplant treatment nonadherence 86–87
precautionary approach 44–46

public engagement 135–136
Public Health Service Act 84

quality of life 61, 63, 71–72
quarantine 83–84

Ramsey, P. 78
Ravelingien, A. 46
Reese, P. 135
Regan, T. 27
regenerative medicine techniques 21
reproductive rights restrictions 82–83
Repo Men (movie) 105–106
research 1–7, 58, 69; economic interests and costs 110–111, 117–120; ethics of 25–30, 94–97, 109–117; future of 124–125; global governance 125; human 59–60; limiting 60–64; objection to 21–22; principles of 30–37; reasons for further genetic engineering 37–38; and zoonosis 44–50
Resnik, D. 115–116
Revivicor 109, 117–118
Reyneke, M. 120
right to withdrawal 51–54
Rollin, B. E. 34
Ross, L. F. 45
Rothblatt, M. 48

Sade, R. M. 43
Sanner, M. A. 133
Saylor, K. W. 64–66, 86
Sharp, L. A. 133
Singer, P. 39n1
social justice 94
Spain 17
Sparrow, R. 48
subject-of-life criterion 27
stigma 133
Swazey, J. P. 70

total organ replacement 19–20

UAB Xenotransplantation Program 112
Ulysses contract 54, 81–82
undue inducement (influence) 78–80
unified brain theories of death 16
United Nations Climate Change Conference 125
United States Constitution 106
United States Department of Health and Human Services 78, 132
United States Public Health Service 82–83
United States Secretary of Health and Human Services 84
United Therapeutics Corporation 112, 117–118
Universal Declaration on Bioethics and Human Rights 135
University of Alabama at Birmingham (UAB) Comprehensive Transplant Institute 112
U.S. Department of Health and Human Services 87–88

Vanderpool, H. 78
Veatch, R. M. 45

waitlists for transplantation 11, *11*, 134–135; mortality 13, *14*, 62–63, 65, 92–93
Warren, C. H. 42
Weijer, C. 85
WHO Global Consultation on Regulatory Requirements for Xenotransplantation Clinical Trials 126–127
withdrawal from clinical trials 81–82; *see also* right to withdrawal
World Health Organization (WHO) 61, 94, 125–128, 135

xenograft biobanks 129
xenotransplant patents 8, 107–109; consequential arguments 110–114; ethics of 110–121; help *vs.* profit 117–120; non-consequential arguments 114–117

zoonotic infection 7, 34, 36, 43–44, 54–55, 81–83, 111–112; biosurveillance 50–54; collective consent 46–50; life-long monitoring for 100–101; precautionary argument 44–46

Kolibri 360 GmbH
Ferdinand-Jühlke-Straße 7
99095 Erfurt, Deutschland
produktsicherheit@kolibri360.de